The World is Triangular

Horst Czichos

The World is Triangular

 Springer

Horst Czichos
University of Applied Sciences
BHT Berlin
Berlin, Germany

ISBN 978-3-030-64212-9   ISBN 978-3-030-64210-5   (eBook)
https://doi.org/10.1007/978-3-030-64210-5

This Springer imprint is published by the registered company Springer Nature Switzerland AG
The registered company address is: Gewerbestrasse 11, 6330 Cham, Switzerland

Foreword of the 1st Edition

I got my first impressions of technology and philosophy during the engineer-internship in a company for counting and calculating machines. *The first adding machine the philosopher Leibniz built,* the trainer explained to us. *Philosophers are concerned actually not with technical things* said my friend Jürgen and described with Plato's theory of ideas an example of philosophical thought:

> *… according to Plato, the objects perceived by our senses are only images of "Ideas", the archetypal patterns of all objects …*

The conversation about Leibniz's calculating machine and Plato's theory of ideas—and the discussions about *Marxism* and *Existentialism* in Berlin in the 1960s—were the beginning of an intensive occupation with philosophical questions.

The engineering education provoked a strong interest in the physical basics of technology, so that besides freelance engineering development work in the optical industry, I studied physics at the Free University of Berlin. Thereby I also learned the analytical philosophy of Wittgenstein. After my doctorate, I worked in research, teaching and technical management.

From the long occupation with technological, physical and philosophical topics—and the experience that things today are mostly "complex" and not "mono-causal" understandable—the idea arose to combine the elementary knowledge from the different areas in a book. Mr. Thomas Zipsner, Springer Vieweg, I thank you for the stimulating and constructive discussions for the realization of this interdisciplinary project and Ms. Imke Zander for the careful editorial support.

The book takes a brief look at the development and the state of knowledge of the three fields and thus wants to contribute to a multidisciplinary understanding of the world.

Berlin, Germany   Horst Czichos
May 2013

Preface

The book gives a short presentation of the triad *philosophy—physics—technology* against the background of the common origin in ancient times.

The emergence of the book has been described in the foreword of the first German edition. The present second edition of the book is updated and extended, whereby new physical research results and technological innovations were included:

– The physics of space and time after the experimental detection of gravitational waves (Nobel Prize for Physics 2017).
– The New *International System of Units* (SI) for Physics and Technology which is completely based on natural constants and entered into force on World Metrology Day, 20 May 2019.
– Actual overview of basic technologies: Material, Energy, Information.
– Technologies for the "Digital World" of information and communication.
– Mechatronic and Cyber-physical systems for Industry 4.0.

The significance of technology for the world in the twenty-first century is discussed in the final section of the book.

Berlin, Germany   Horst Czichos
July 2020

Contents

Chapter 1
The World of Antiquity

The origins of philosophy, physics and technology are embedded in the world of antiquity, which is considered with its facets in the first part of the book.

In the cultural *Axial Age* (Karl Jaspers), in the period from 800 to 200 B.C., in several cultural areas the philosophical and technological developments evolved, which still form the basis of all civilizations today, Fig. 1.1. Even before that, about the 2nd millennium B.C., the "religions of revelation" (Judaism, Christianity, Islam) originated, who refer to Abraham as their progenitor. At the same time, the Phoenicians (Lebanon, Syria) developed the easily learnable alphabetical script, from which the European writings (Greek, Latin, Cyrillic) descendants.

1.1 Ethics—Religion—Natural Philosophy

In Persia, Zarathustra (ca. 630–550 B.C.) taught that people could choose between have good or evil. Good virtues are good disposition, truthfulness, wisdom, dominion, health, longevity. Evil is deceit and anger. A principle of life is the Triad: *good thoughts—good words—good deeds.* The world is the place of struggle between the good and the bad, in the end the spirit of good will prevail.

In China, Lao-tse (604–520 B.C.) cites the term *Tao* as the origin of the world, which embodies the unity and harmony of all things. Man should recognize this principle of originality, naturalness and simplicity as "way" and orient thinking and acting align with it.

The hallmark of the perfect human is silence, a philosophical non-action, the refusal to intervene in the natural course of things. State and rule should be kept to a minimum. The more laws and regulations, the more lawbreakers there are.

Confucius (551–478 B.C.) developed the principles of humanity and reciprocity, which should have a balancing and harmonizing effect in society in order to prevent injustice. The central concern is the embedding of the individual in family, State and morality. The five relationships, Prince ↔ Civil servant, father ↔ son, man

© The Author(s), under exclusive license to Springer Nature Switzerland AG 2021
H. Czichos, *The World is Triangular*,
https://doi.org/10.1007/978-3-030-64210-5_1

Fig. 1.1 Cultural areas in which the foundations of civilizations arose

↔ woman, older brother ↔ younger brother, friend ↔ friend, must be determined through humanity, right action, custom, knowledge, truthfulness.

In India, the historical Buddha, Siddhartha Gautama (560–480 B.C.) founded the teaching of Buddhism. The teaching is based on the four noble truths: (1) everything life is full of suffering; (2) the cause of suffering is "thirst", desire; (3) suffering can be overcome by the suppression of desires and passions; (4) the path to this is the *noble eightfold path* with the steps of wisdom, morality, deepening. The goal is healing, the suspension of the ego-centered existence, the extinction of the life illusions, the nirvana.

Symbols of eastern cultures are shown in Fig. 1.2.

All cultures emphasize the importance of Ethics.

Golden rules of practical ethics

- What should not be done or forced to me I will also not do to other humans (Confucius)
- What is unwelcome and awkward to me how could I do it to others (Buddha)
- Nothing doing to other humans which is not good to myself (Zarathustra)

The golden rules of ethics are comparable with the biblical commandment of charity, which according to the Luther Bible of 1545 as a Christian rule of life is expressed in this way:

- **What you don't want done to you, don't do to others.**

Under the term **Religion**, a multiplicity of different cultural phenomena that are related to elementary realities of life (birth, body, soul, death) are summarized. T*heism* is the conviction of the existence of a (personal) God, which is in distinction to the counter-concepts of *atheism* or *pantheism* (God and nature are one).

However, science can neither prove nor disprove the existence of God. *We do not know what the meaning of life is and what the right moral values are. A discussion*

Fig. 1.2 Symbols of Eastern Philosophy: Tori (left), symbol of Japanese Shintoism and Buddha statue

of this necessarily leads to the broad source of attempts at interpretation and moral teachings and these fall within the realm of religion, says the Nobel Prize winner Richard P. Feynman in his book *The Meaning of it All* (Feynman 1999), and named three basic aspects of religious belief:

- the metaphysical aspect, which tries to explain what things are and from where they come, what man is, what God is and what qualities he has,
- the ethical aspect, which gives instructions on how to behave in general and especially in the moral sense,
- the inspiring power of religion for art and other human activities.

Today, there are more than 2 billion Christians, 1.2 billion Muslims, 810 million Hindus, 380 million Taoists, 360 million Buddhists, 12 million Jews and 6 million Confucians on earth, as well as millions of believers who subscribe to other theories and teachings (Anke Fischer in: The seven world religions).

- *Judaism, Christianity, Islam* have their roots in the "Abrahamic model of divine revelation", according to which the world was created by a benevolent God. The realities created by God in the physical world are worth less than the humans they are subject to. Man should therefore not be guided by reality of this world, but must seek his model of behavior in God himself. God reveals himself less through his creation than through his "revelation". He can proclaim a "law" to the world via "prophets" or "angels", as in Judaism (Torah) and in Islam (Koran), or through the incarnation as "Son of God" to enter the world as in Christianity. Christianity emphasizes the principle of charity and proclaims that the soul is considered a personal attribute of every human it is immortal, but indissoluble with a single individual connected.

- *Buddhism* is a non-theistic religion that is directed at all people. It does not understand itself as a revelation, but as a discovery of the world connections with the four noble truths and the noble eightfold path.
- *Hinduism* knows no founder of religion and no dogma, thinks cyclically (birth, death, rebirth) and is based on the religious-philosophical System of Sanatana Dharma (eternal law) with different concepts of God and the principle of karma (transmigration of souls).
- *Confucianism* developed out of the teachings of Confucius and was still in use until at the end of the Chinese Empire as temporarily Chinese state religion. It is based on the principles of humanity and reciprocity.
- *Taoism* (also known as Daoism) is based on the teachings of Lao-tse and is concerned with the harmony of man and nature. At the center of the religious Taoism stands above all the search for immortality.

The **Natural Philosophy** of the ancient world embraced—with a pantheism view—nature and the soul and actions of mankind. It has its origins on the west coast of Asia Minor (Ionia, today Turkey), on Sicily, in Lower Italy and in Athens, Fig. 1.3. "We are attracted to Greek philosophy *because nowhere else in the world, neither in before and after, such an advanced, well-structured building of knowledge and thought has been established,*" emphasizes the physics Nobel laureate and co-founder quantum physics Erwin Schrödinger in his book *Nature and the Greek* (Schrödinger 1958).

Thales of Miletus (ca. 625–547 B.C.) is considered the first philosopher ever. As a mathematician, he discovered the theorem of Thales (every triangle enrolled in a hemicycle is right-angled) and is said to have calculated the height of Egyptian pyramids with the mathematical theorem of rays. As a philosopher he took a universal invigorating principle, literally translated "water", as symbol for the diversity of

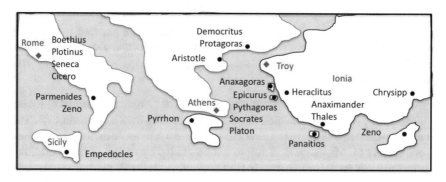

Fig. 1.3 Philosophers of Western Antiquity and their birthplace or location of activity

dynamic, life-sustaining functions. Anaximander (611–545 B.C). and Anaximenes (585–525 B.C.) from Miletus tried to trace the "All of Being" back to a first principle (ark). For Anaximander it was the principle "infinity" (wherefrom things arise and where they go). For Anaximenes the first principle was "air" because "air carries the world".

Heraclitus of Ephesus (544–483 B.C.) developed the concept of *Logos* as cosmological, all-founding and determining principle and raised the "fire" to its symbol and first principle (arche). The "embeddedness" of physics in the philo-sophia of antiquity was revealed by the physicist Heisenberg in his book Physics and Philosophy (Heisenberg 2011) with the following comparison:

If one replaces the word "fire" with the word "energy", one can call Heraclitus' statements as an expression of our modern physics. The energy is actually the substance from which all elementary particles, all atoms and all things in general are made, and at the same time, the energy is also the moving thing.

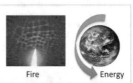

Fire　　Energy

Heraclitus assumed that everything is determined by "change" as a structural principle. He illustrates the constant change through the image of the river, whose water is constantly changing and yet remains the same. (You can't go down the same river twice). Heraclitus' principle of constant change was later expressed by the aphorism "everything flows". Figure 1.4 shows analogies between eastern and western symbols.

Parmenides of Elea (515–445 B.C.) developed the basic features of a "doctrine of the knowledge of being". The being is un-become and imperishable. It is an "indecomposable Whole" and must be regarded as uniform and dormant: "The same thing namely is knowledge and being". The followers of the doctrine of Parmenides are called eleates. Zenon of Elea (490–430 BC) defended the doctrine of Parmenides by perceptive and convincing (dialectical) evidence:

Anaxagoras of Klazomenai (ca. 500–428 BC) was one of the first philosophers in Athens. He started from the principle that "nothing can be created from nothing" and that there is an infinite variety of material qualities in small units. According to today's interpretation, his considerations can be understood as anticipation of the principle of self-organization of matter. Anaxagoras saw—in contrast to the mythical-magical thinking of his time – e.g. in the stars no divine beings, but celestial bodies similar of the earth. The views of Anaxagoras gave natural philosophy expanded dimensions.

Tao is the origin of the world, which embodies the unity and harmony of all things, symbolized by Yin (white, cold, negative) and Yang (black, warm, positive). The S-shaped curve symbolizes the dynamic character of dichotomy. Lao-tse

Everything is governed by Logos, the law of the unity of thesis and antithesis. Nothing is imaginable without its antipode: day and night, vigil and sleep, life and death. All events result from contradictions. Change is the general structural principle.　　Heraclitus

Fig. 1.4 Symbols of Eastern Philosophy (Lao-tse) and Western Philosophy (Heraclitus (

1.2   Primordial Elements and Atomism

After Thales had declared "water" the very first beginning (ark), Anaximenes the "air"
and Heraclitus the "fire", Empedocles (495–435 B.C.) was first thinker of antiquity
to extended this "monism" to a "pluralism". He designated the four elements *fire,
earth, air, water* as the "roots of things". Empedocles supposed in mythological
language, that through "love" and "discord" between basic elements the variety of
things comes into being.

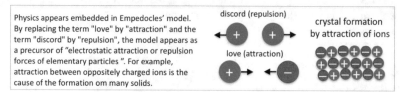

Physics appears embedded in Empedocles' model.
By replacing the term "love" by "attraction" and the
term "discord" by "repulsion", the model appears as
a precursor of "electrostatic attraction or repulsion
forces of elementary particles ". For example,
attraction between oppositely charged ions is the
cause of the formation om many solids.

Aristotle (384–322 B.C.) supplemented the model of Empedocles with a fifth
primordial element (quintessence) as a symbol of heaven (sun, moon, stars). The five
elements were symbolized by perfectly regular polyhedrons, Fig. 1.5. The symbolic
bodies can be disassembled and new "model building blocks of matter" can be built
from them. For example, a tetrahedron (symbol fire) and two octahedrons (symbol air)
can be disassembled into twenty equilateral triangles and from these an icosahedron
(symbol water) can be built up.

In other cultures, similar models of the "elements of nature" were developed—in
the Buddhism the four-element doctrine and in Taoism the five-element doctrine,
Fig. 1.6. The idea that nature in all its manifestations is made of a special mixture of
the four basic elements was explainable until the beginning of the nineteenth century,
i.e. until the discovery of oxygen and its subsequent formation of today's chemistry,
scientific claim.

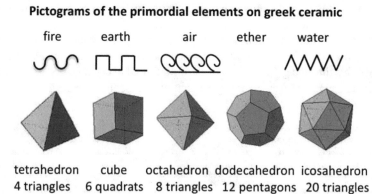

Pictograms of the primordial elements on greek ceramic

| fire | earth | air | ether | water |

| tetrahedron | cube | octahedron | dodecahedron | icosahedron |
| 4 triangles | 6 quadrats | 8 triangles | 12 pentagons | 20 triangles |

Perfectly regular polyhedra symbolizing the primordial elements

Fig. 1.5   The primordial elements of antiquity and their symbols

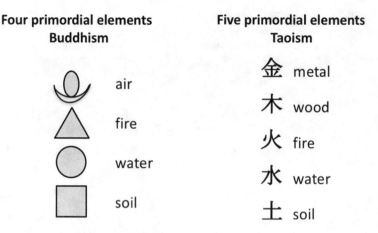

Fig. 1.6 The primordial elements in Eastern cultures

1.2.1   Atomism

Ancient atomism—founded by Leucippus of Miletus (460–370 B.C.) and Democritus of Abdera (460–371 B.C.)—postulates the existence of smallest (even indivisible) particles (atoms), which in different combinations determine the type, shape and change of things. All atoms have their own original capacity to move. Between the atoms there is the "void", the space between them. The whole reality can be completely defined in its structures and processes by the different types of movement, the connection or dissolution of atomic groups in the void. Atomism established a completely new understanding as illustrated in a comparison with the earlier "Natural Philosophy", see Fig. 1.7.

- The philosophy of nature of Thales is looking for a uniform "primordial substance" and takes archetypal "water" as the primordial element.
- In contrast, ancient atomism postulates that all things—including the primordial element "water" of Thales—consists of smallest particles, the atoms.
- The atomic structure of matter postulated by Democritus around 400 BC could be made "visible" experimentally in 1951 with field ion microscopy, see Sect. 3.1.

Using the model of atomism, Democritus explains with the formal characteristics of atoms the differences between the four elements postulated by Empedocles (earth, air, water, fire) and the fifth element, the ether, that substance that according to Aristotle forms the celestial world, thinking and intelligence.

Also the spirit is made of atoms, it is a "psychic matter". This statement is the first philosophical representation of "materialism", according to which everything that exists can be explained without exception as a combination of basic atomic types.

In the theory of atomism, a certain analogy to graphic texture can be seen because atoms combine with each other to matter like alphabetic characters combine to words. So, the things and beings of the world are dependent on:

- **Natural Philosophy of Thales** (ca. 625-547 B.C.)
  Archetypal model: "water": symbol of the diversity of life-sustaining functions

    water as liquid   water as vapor (cloud)   water as solid (ice)

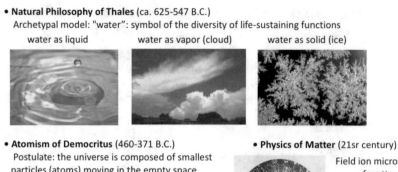

- **Atomism of Democritus** (460-371 B.C.)
  Postulate: the universe is composed of smallest
  particles (atoms) moving in the empty space

     • • • • • matter (atoms) exists
     • • • • • emptiness exists
     • • • • • the cosmos is infinite
     • • • • • a central point does not exist
     • • • • • the physical laws are universal

- **Physics of Matter** (21sr century)

Field ion microscopy
of matter:
The bright spots can
be interpreted as
images of single
atoms,
see Fig. 3.1

Fig. 1.7 Models of Natural Philosophy, Atomism and the real physics of matter

- the shape of atoms from which things are formed differ, like the shape of letters from which words are formed differ, e.g. letter A has a different shape than letter N,
- the position of atoms in a "composite" (substance) or of the position of letters in a "phrase",
- the order of atomistic or letter combinations. A limited number of atoms explain the complexity of the world, similar to twenty-six letters of the alphabet are sufficient to form all words.

1.2.2   Determinism

The philosophical and scientific determinism established by Democritus claimed that there is a necessary relationship between all natural phenomena which is based on the principle of cause and effect.

A system is described as deterministic if by its "state" at an arbitrary date and applicable "laws of nature" the state of the system to any future date is completely determined. If the laws of nature assumed here are thought to be "causal laws", one speaks of *causal determinism,* or with respect to the world of a *cosmological determinism.* A special case of the causal determinism is "theological determinism" with the "doctrine of predestination", that all events in the world are predestined by God's (inexplorable) fact and will.

1.3 Measure and Number

For the philosophers of classical antiquity, **measure** is a basic concept that defines the "coherent". A life lived in harmony can be described as something that is "ordered by measure".

1.3.1 Rules of Governance

Nothing in excess is a well-known aphorism ascribed to Solon (640–560 B.C.), the first historically proven creator of the law of the Athenian city-state. The wording of Solon's laws, in the first "Athenian Constitution" is not preserved. To the question of the best state, Solon should have answered (conveyed by Plutarch's document *The Banquet of the Seven Wise Men)*: "The state, in whom a criminal is accused and punished in the same way by all those he has not harmed and the one he has harmed".

The Greek philosopher and politician Demetrios of Phaleron (350–280 B.C.) ascribes the following aphorisms to Solon:

- Do not advise the citizens what is most pleasant, but what is best.
- If you demand accountability from others, give it yourself.
- Be more faithful to decency than to your oath.
- Do not sit in judgment, or you will be an enemy to the condemned.
- Learn to be ruled, and you will know how to rule.
- Lie not, but speak the truth.
- Seal your words with silence, your silence with the right moment.
- Unlock the invisible from the visible.

The saying "the measure is the best thing", which comes from classical antiquity can be seen as a feature of Greek thinking as a whole. Aristotle put it this way:

The true man chooses the medium and distances himself from the extremes, the excess and the too little.

1.3.2 Number Symbolism

Number is an abstract functional unit of account in mathematics and should not be confused with the character that symbolizes a number, e.g. "4" and "IV" are different symbols for a single number.

Pythagoras of Samos (around 580–496 B.C.) founded the religious-philosophical movement of Pythagoreism, a movement based on numbers teaching. Since Pythagoras, the natural (integers) numbers are in the center of occidental interpretation of nature. "*I don't know any other person who has had such an influence on human thinking like Pythagoras,*" says the philosopher Bertrand Russel (1872–1970).

Philolaus (about 470–400 B.C.) wrote the first script of the Pythagoreans and took the view that "numerical ratios" did not only represent the stellar system (with the earth as movable celestial body), but also arrangements of concrete things and moral values in a symbolic way. Pythagoras connects with the aphorism "*everything is number*", the idea that the cosmos is defined by numerical proportions which, from a philosophical point of view, are called "harmonies of the spheres".

He had recognized that the harmonic harmony of sounds can be achieved by numerical proportions and discovered that a monochord with a string over a movable bridge, at a musical octave the string length ratio 1:2, for a fifth is 3:2 and for a fourth is 4:3, Fig. 1.8.

The number symbolism of the Pythagoreans assumes a "magic power of numbers", Fig. 1.9.

The integer number are assumed to have the following symbolic meaning:

– The number **1** is the "creator" of the numbers 1, 2, 3.
– The number **2** marks "duality" as the middle between unity and trinity, it can also have a negative meaning (double-dealing).

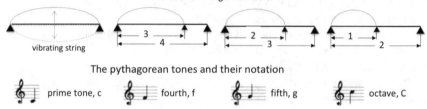

Pythagoras observed that by an an integer subdivision of a vibrating string a correlation between sound and number is possible. He developed the foundation of music theory representable by a relation of integer numbers.

Fig. 1.8 Fundamentals of music theory according to Pythagoras

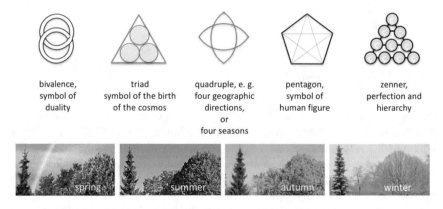

Fig. 1.9 Illustrations of the Pythagoreans' symbolism of integer number

- The number **3**, as triple step of *Being–Living–Thinking*, is the symbol of the cosmic birth. Various doctrines know the term "divine triad".
- The number **4** is hypothetically associated with the "created". The four body-states *cold, warm, humid, dry* are considered to be the basic concepts of medicine and a psychology of "temperaments" constructed from it:

1. Phlegmatic (cold-damp),
2. Choleric (warm-dry),
3. Sanguine type (warm-humid),
4. Melancholic (cold-dry).

There are the four lifetimes *childhood, youth, maturity, old age* and the four seasons *spring, summer, autumn and winter.*

- In the ancient tradition, the **5** links the pentagon with the figure of man (head, two hands and two feet).
- The **7** is according to Pythagoras "the perfect number". It combines the duality of mind and soul and the pentagon symbol of man.
- The number **10**, sum of the numbers 1, 2, 3, 4 (today the basis of the decimal system) was considered by the Pythagoreans as the "mother of all numbers".

Pythagoras developed an "arithmetic geometry" and related numbers besides their abstract mathematical dimension also to a spatial significance. The according to him named "Theorem of Pythagoras" is used in mathematics lessons in every school (Fig. 1.10).

(Also taught at school since ancient times are the theorems of geometry of Euclid of Alexandria (360–280 B.C.). The Euclidean geometry is an axiomatically constructed doctrine of the descriptive geometry of the plane or the three-dimensional space.)

The Pythagorean theory of the "descriptiveness of the world by natural numbers" already reached its limits in the analysis of a simple square with side length 1, the

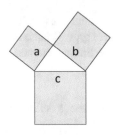

Pythagoras' theorem states that in a right triangle the area of the square whose side c is the hypotenuse (the side opposite the right angle) is equal to the sum of the areas of the squares on the other two sides a and b.

$$a^2 + b^2 = c^2$$

Incommensurability
The relation between the side length and the length of the diagonal of a square can not be expressed as ratio of two integer numbers. The number $\sqrt{2}$ is irrational.

Fig. 1.10 Pythagoras' theorem and the Incommensurability

diagonal of which is 'incommensurable', i.e. its length cannot be expressed as an integer or as a ratio of integers, but is an "irrational number", namely $\sqrt{2}$. The discovery of the incommensurability—initially kept secret by the Pythagoreans—shock the validity of the Pythagorean model of number symbolism. A mathematically correct definition of irrational numbers was first achieved by Georg Cantor and Richard Dedekind in the late nineteenth century.

1.4   Foundation of Philosophy

The early thinkers of natural philosophy tried to create a model of the world as a well-ordered whole thing. When it became clear that the individual different hypotheses of "reductionism to a single original principle" could the world not universally explain, broader philosophical views were developed.

1.4.1   Moral Philosophy: Socrates

Socrates of Athens (470–399 B.C.) founded the classical period of the Greek philosophy and makes the turn from the "pre-Socratic" natural philosophy to ethics. He turned against the elder motto of believing only what one would like to, because that leads to the "dissolution of the idea of truth". His criticism was directed in particular against the mixing of philosophy and the spectacle of the rhetorical representation of antinomies (contradictions) in which first a thesis is presented with powerful arguments and then refuted with equally valid arguments - thus the controversial nature of each thesis could be demonstrated. Socrates trusts only of reason, whose inherent lawfulness in reasonable conversation leads to true insight. He compares his activity with that of his mother, a Midwife. Just as she helps a mother giving birth, Socrates helps his interlocutor at the "birth of truth".

Socrates established the method of trying to get at truth by persistent questioning. At the beginning of a Socratic dialogue there is usually a question of type: What is X? The Variable X stands for concrete ethical issues. The dialogue technique pursued a double objective:

– the critical examination and/or refutation of assertions,
– the generation of a reasoned answer to the initial question by means of a definition of terms.

Socrates' reason-based, myth and superstition rejecting discussions with the Athenian citizens were misunderstood by the authorities. He was arrested for impiety and seduction of youth sentenced to death (drinking hemlock cups). Socrates left nothing in writing. An example of Socratic Dialogues gives his student Plato in an impressive way. The knowledge sought by Socrates is a practical knowledge, which has the knowledge of good and evil as its content, it leads to critical self-examination and provides people with a rational approach to their actions.

The philosophical activities of Socrates led to the most important philosophical developments in human history:

- the Dual Thinking of Plato, student of Socrates,
- the Metaphysics of Aristotle, student of Plato.

1.4.2 Dual Thinking: Plato

The "dual thinking" founded by Plato (428/427–348/347 B.C.) is an extension of the "monism" of natural philosophy, which searched for a single universal principle of order of the world and of thought. Plato's dual thinking establishes a fundamental extension: philosophical thinking is no longer directed only towards the "reality", but becomes a reflection on the "knowledge of reality". In the fundamental work of his philosophy, the Timaios, Plato describes a World view that demands a certain image of man. Platonism makes "the good" to the supreme principle. The good exercises its dominion over physical reality, it regulates the morality of the soul and gives the state in which humanity is developed the unit without which it would go down. Plato discusses the difference between.

(a) an ever-changing world of phenomena, according to Heraclitus' aphorism "everything flows",
(b) the world of always constant and unchangeable beings, according to Parmenides.

Plato's philosophy divides the world, so to speak, into a "realm of changing phenomena" and a "kingdom of eternal being". He considers the contrast between a "concrete single thing" that comes into being and passes away, and the "idea" behind it as its "principle". To explain these differences, Plato leads in his work Politeia the so-called parable of lines, Fig. 1.11. Everything we refer to in this world is represented by a line with a dual division into.

(a) accessible to thinking → **knowledge**: ideas, mathematics, reason (supreme cognitive faculty), mind (lower, with notional thinking connected cognitive faculty,

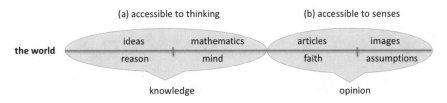

Fig. 1.11 The world in the model of dual thinking according to Plato

(b) accessible to the senses → **opinion**: articles, images, faith, assumptions.

By "ideas", Plato understood the highest and unchangeable things (entities), which dictate in the last instance all to be, to recognize and to act. Examples are "the good", "the true," "the beautiful". The sensually perceptible objects of the world around us are merely a "partner" in the ideas. As Plato points out in his "Parable of the Cave", only "shadows" (images) of "ideas" (archetypes) are accessible to the senses. The ideas that exist outside our perceptual world are "archetypal patterns" of all sensual objects. According to Plato's philosophy, the ideas must be the base to orientate our thinking and acting, if we want to find the lasting and essential in all things understand or want.

In today's understanding, however, "ideas" are understood only as "subjective entities", which differ from person to person. *"A more modern term for what Plato calls the idea is structure or form. Matter in the sense of the classical physics of the occidental modern age does not exist for Plato"* (Carl-Friedrich von Weizsäcker).

In his theory of ideas, Plato confronts the self (the "I", the "subject") and the "being" of real objects with a third: the "absolute" of ideas in the mind. According to the philosophy of Plato, ideas are eternal and unchanging. They are "the pure truths of themselves" and under the terms of "the good and the beautiful" they are a symbol of the divine. Plato's theory can be symbolized by the **Platonic triangle**, Fig. 1.12.

The school of philosophy founded by Plato, the Platonic Academy in Athens, discontinued its teaching activities at the end of the 80th BC as a result of the Roman civil wars. The philosophers of the post-academic period are now called Middle Platonists. An important site of central Platonic activity was Alexandria, a port city on the Mediterranean coast of Egypt, which is home to the most important library of classical antiquity an economic, spiritual and political center of the Roman Hellenistic. Neo-Platonism—the philosophical current prevailing in late antiquity—was founded by Plotinus (204–270), a philosopher educated in Alexandria. The most important extension compared to the Platonic doctrine is the assumption, that all reality can be traced back to a "primal One" as its origin. Plotinus equates the "primal One" with the "good" and thus gave it the function of a universal positive principle.

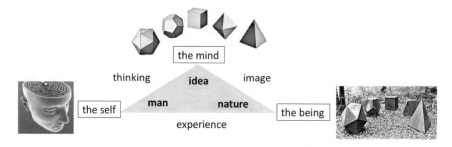

Fig. 1.12 The Platonic triangle, symbol of the theory of ideas

1.4.3 Metaphysics: Aristotle

Metaphysics formed from Greek words *meta* (after, beyond) and *physics* (natural history) is a discipline of philosophy founded by Aristotle (384–322 B.C), see Sect. 2.2 *Philosophy of being*.

Metaphysics asks for the elementary reasons why.

(1) "something appears as something" at all, and
(2) "as such" is recognizable.

The central concepts of Aristotelian metaphysics include potential (possibility, capability) and actuality (reality). For example, the seed potentially contains the plant, and the actual plant is nothing other than the development of the plants contained in the seed. The concept of the Aristotelian metaphysics is diametrically opposite the concept of Plato's doctrine of ideas. According to Aristotle, "being" is not an "idea" but the concrete "first substance" (substantia prima), a sensually perceivable "thing" from animate or inanimate nature or from the world of artistic or technical things. The platonic "idea" on the other hand is something supernatural, spiritual, with which we try to establish our sensual spatial–temporal world. Aristotle thinks the other way around: First, the spatial–temporal world is there and this concrete, nothing else, means "to be" in the true sense. In a simplified comparative consideration of the two views, the Platonic model can be seen as an epistemological idealism and the Aristotelian model can be considered epistemological realism.

The aim of the school of philosophy founded by Aristotle was to create an **encyclopedia of knowledge**. In this process, the terms that are still valid today were found for the various fields of knowledge, including *logic, physics, meteorology, economics, rhetoric, ethics, psychology* and many terms that are still used today, such as *energy, dynamics, induction, substance, category, attribute, universality*. Each science has its own qualitatively different subject and is based on its own principles. Mathematics studies the phenomena produced by "quantity". Physics looks at things that have a physis (body) and investigates the causes of the "movement" of bodies. Biology is interested in the problems of "life". Theology is the knowledge that deals with the "divine" and astronomy studies the manifestations of the stars of the "sky". There are no more or less important sciences. All disciplines *have* the same dignity, since they interpret nature from different points of view.

According to Aristotle, there are four basic characteristics (Latin *causa)* of all man-made things:

- Substance: what-is-it-made-of (*causa materiali*s):
- Form: *what-is-the-shape-by-which-it-is-identified (causa formalis)*
- Effectiveness: what-actually-does-or-make-it (*causa* efficiens)
- Purpose: *the-ultimate-reason-for-it (causa finalis*)

Things are made by *techne* (Greek craftsmanship) to achieve their purpose in space and time. The general principle to characterize things is depicted in Fig. 1.13a together with a classic example.

(a)

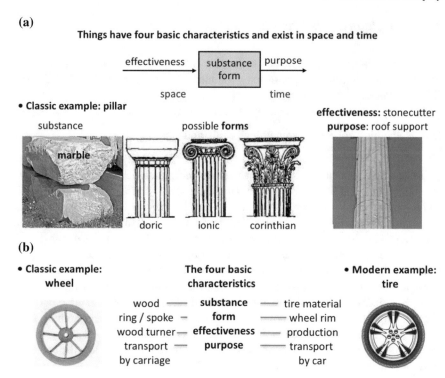

(b)

Fig. 1.13 **a** The basic characteristics of things and a classic example. **b** A classic and a modern example of the basic characteristics of things

The theorem of Aristotle that all man-made things have four basic characteristics is of universal validity and applies also to things produced by modern technology. An example shows Fig. 1.13b.

1.5 The Cosmos

"The word cosmos was introduced by Plato in his work Timaios (ca. 360 B.C.) in the meaning of world into the history of language—as the first description of reality, which forms an ordered whole that is both good and beautiful", writes the French philosopher Remi Brague in his book *The Wisdom of the World- Cosmos and World Experience in Western Thinking* (Brague 2006). The ancient model of the cosmos was developed, with the help of Aristotelian theories of the second century A.D., by Claudius Ptolemy (100–180). Already earlier, Aristarchus of Samos (310–230 B.C.) presented the first known *heliocentric model* and placed the Sun at the center of the known universe with the Earth revolving around it. Ptolemy saw no reason why the earth should rotate around the sun because then a "parallax", i.e. a shift of

Fig. 1.14 Model of the antique cosmos

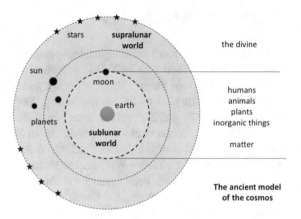

the planetary constellation could be observed against the background of the fixed stars. (The argumentation of Ptolemy is theoretically correct, but the real existing parallax is not visible to the naked eye and can only be seen with high resolution astronomical instruments).

According to Ptolemy, the earth is firmly at the center of the universe. All others Celestial bodies (moon, sun, the planets and the starry sky) move on orbits around this center. This geocentric world view describes the cosmos as a structure of concentric layers which are divided into two zones with different laws. Figure 1.14 gives an overview of the cosmos model.

- The *sublunar world* includes as center the earth with transient matter, plants, animals and humans and extends with different zones (water, air, ether) to the sphere of the moon.
- The *supralunar world* reaches to the border of the universe and knows no change, because it is divine in nature. The sun, the moon and all celestial bodies are a manifestation of the unseen divine.

Aristotle imagines the order of the cosmos in a stepped mental arrangement. According to the terms "potential" (possibility, capability) and "actuality" (reality), matter is only potential, while the beings formed from matter have potential and actuality. The highest value is the "divine". The divine is substance-free topicality and forms the highest point of the hierarchical order of all beings. Aristotle coined the expression "unmoving mover", since the divine as the highest power moves everything in the world, but remains immovably unmovable itself. The "highest concept" of the divine was given by Aristotle in his systematically structured philosophy as follows: Since all things order according to the degree of their value (e.g. smaller, bigger or good, better), there must also be a general conclusion in terms of value, since an increase is excluded to infinity. Aristotle's metaphysical doctrine of God plays as "cosmological proof of God" in medieval philosophy and theology an important role.

1.6   Culture and Art

The cultural history of antiquity begins, according to the classical view, with Homer, whose vita in the period 1200–400 B.C. is historically not exactly documented. He is called the creator of the world literature epics *Iliad* and *Odyssey* and is the first and most important poet of the ancient world. The Iliad (German translation by Heinrich Voss, 1779–1822) describes a section of the Trojan War. Troy (discovered 1873 by Heinrich Schliemann) was on the 15 m high settlement hill Hisarlık at the Dardanelles and controlled since the Bronze Age, the access to the Black Sea. Mythical trigger of the Trojan War was the abduction of Helen, wife of Menelaus (King of Sparta), through Paris (son of the Trojan king Priam). After ten years of siege by united Greek warriors, the Greek conquered the strongly fortified city by means of the stratagem of the "Trojan horse" of Odysseus—a wooden horse with Greek warriors in interior—which was brought into the city by the Trojans. The epic Odyssey describes the adventures of king Odysseus of Ithaca and his companions on their homecoming from the Trojan War. The term "Odyssey" became a symbolic term for long wanderings. The Homeric epics have influenced European art and culture to the present day humanities. They have contributed significantly to the development of ancient religious ideas and describe the gods of Olympus. The main gods are: *Zeus* as the father of the gods and family goddess *Hera*, *Poseidon* god of the sea, *Apollo* god of light, *Athena*, goddess of science, *Aphrodite* goddess of love, *Hermes* messenger of the gods. The work of Homer illustrates as in a kaleidoscope (Greek: beautiful forms see) the fundamental foundations of the Greek spirit: the idea of divine, the cult of hospitality, individual courage, love, joy the beauty and close observation of nature.

The four "cardinal virtues" are moderation, wisdom, justice and courage. The conceptualization goes back to Aeschylus (525–456 B.C.), who, with Sophocles (497–406 B.C.) and Euripides (486–406 B.C.) are among the great tragedians of antiquity. Tragedy is a unique and extraordinary creation of the Greek spirit, it still influences the theatre world today. The Greek tragedy deals with fateful entanglement in a hopeless situation, characterized by the attribute "guiltless guilty". The themes treated by the Greek tragedians range from philosophical to religious and existential questions. The Fate (or, the gods) put the actor in an indissoluble situation—a conflict typical for the Greek tragedy—which caused the inner and outer collapse of the person. There's no way not to be guilty without abandon his values. In the classical performances, the choir (placed as dialogue partner in the orchestra in front of the mostly three actors) has the task—as "voice of outside"—to comment and to interpret the performance politically, philosophically or morally. The Dionysus Theatre (southern slope of the Athenian Acropolis, Fig. 1.15) is considered to be the birthplace of tragedy and theater in Greek antiquity. At this place, every year the Dionysus Festival was held to honor with singing, dancing and sacrificial rites, the god of wine and ecstasy.

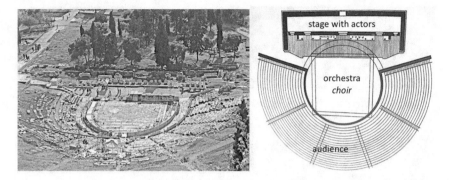

Fig. 1.15 Dionysus Theatre, built about 330 A.D. with 78 tiers for an audience of 17,000 viewers

1.6.1 The Role of Proportions

Thinking in proportions was one of the most used methods in culture and art. Artists and architects developed special proportionality rules for works of art and buildings. Proportion is an equivalence ratio for the description of circumstances—designated with A, B, C, D—which are related. The proportionality rule states that A behaves to B, like C to D, mathematically expressed: $A/B = C/D$. With regard to the application of thinking in proportions in art, the question is whether visual beauty can also be expressed through numerical relationships. Various theories have been developed which, with the help of geometry, should explain the pleasure in aesthetic sensibility. Of particular importance were the as "divine proportion" respected "golden section" and the application of proportionality rules in architecture and the representation of the human body, Fig. 1.16.

1.7 Schools of Thought

The ancient philosophers understood philosophy as a system of knowledge with the components logic, ethics and natural philosophy with various directions of thought.

1.7.1 Sophists

In the period 450–380 B.C., in the Greek city-states the new influential intellectual movement of the Sophists is known. They made thinking itself an object of reflection. The Sophists were itinerant speakers, educators and political advisors, and they mediated for a fee, knowledge about all areas of practical life, whereby they took a critical stance towards traditional beliefs. Their dialectical argumentation technique

Art of antiquity with classic proportions

Golden ratio: proportionality of a and b, such that a:b = (a+b):a = Φ, Φ = (1 - √5)/2 ≈ 1,618	a	b	Goddess of grace (Metropolitan Museum of Art, New York)
	a	a	
	61,8 %	38,2 %	

Girl statues (Karyatiden) at the temple of Acropolis in Athens

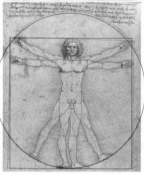

Venus de Milo with classic proportions (Louvre, Paris)

Proportionality study of the human body by Leonardo da Vinci (1492) after a model by Vitruvius (1st century B. C.). The ratio of quadrat side length to circle radius corresponds with a deviation of about 2% to the golden ratio

Proportionality rules of Polykleitos for statues

The rules of Polykleitos recommend to give statues a dynamic equilibrium through a well-balanced proportional relation of the body parts to each other.

• The bended left leg and the backed food should correspond to a lowering of the right shoulder.
• The bended left leg should be allocated by a flexed arm on the same side.
• The supporting leg corresponds to the downcast right arm.
• The head (1/8 of the body) should bow to the opposite side of the bended left leg and the flexed arm

Statue by Polykleitos Roman copy, marble

Bronze statues (2 m) found 1972 in Riace, allocated to Polykleitos

Fig. 1.16 The term proportionality and its application in ancient ar

(art of contradiction) established the relativity of all common statements, principles and terms. The main representatives of Sophism were Protagoras of Abdera (ca. 485–415 B.C.) and Gorgias of Leontinoi (ca. 485–396 B.C.), they advocated the following theses:

• Reality is not directly visible (phenomenalism).
• Every cognition is dependent on the subject (subjectivism, relativism).
• It is impossible to decide about the truth or falsity of any thesis (skepticism).

Fig. 1.17 Controversial dispute in the moonlight (Victor Bogdanov, Saint Petersburg, 1998)

These theses were summarized by Protagoras in the famous saying "Man is the measure of all things". According to Protagoras, this motto expresses the principle of the relativity of all truth claims. Only in relation to certain people one can decide on the validity of claims. The Sophists taught that for every issue, there could be two opposing opinions. The views of the opinion representatives can be "orthogonal" to each other, as a contemporary painting illustrates, Fig. 1.17.

1.7.2 Cynicism

Cynicism is a school of thought in ancient Greek. For the Cynics, the purpose of life is to live in agreement with nature. As reasoning creatures, people can gain happiness by rigorous training and by living in a way which is natural for themselves, rejecting all conventional desires for wealth, power, and fame. Instead, they were to lead a simple life free from all possessions. The direction of thinking of cynicism, also called autarchy, was established by Antisthenes (445–365 B.C.) and Diogenes (413–323 B.C.). The cynics demonstrated frugality ("Diogenes in the barrel") and announced that they have no government, no private property, no more marriage and no more religion. The autarchy developed a pattern of human behavior, which over the centuries and until today has been typical for countless movements (anarchists, dropouts). Autarchy maintains an eccentric behavior and the challenge of the existing order for ethically valuable and interpreted the "freedom" as elimination of superfluous needs and "return to nature". In the Greek-Roman antiquity further schools of thought and various Schools of philosophy arose that gave instructions for action in daily life.

1.7.3   Epicurus

According to ancient tradition, Epicurus (342/341–271/270 B.C.) believed that philosophy is an activity which through thought and discussion creates a happy life. Happiness means health of the body, peace of mind and a fearless quiet life through the conscious elimination of fears, pain and false hopes desires. He stressed that the state of happiness can only be achieved through a virtuous life, sober thinking and secure knowledge. The ethics of Epicurus is called hedonism (Greek hedone: pleasure). It judges against the fear of the gods, the fear of the future, the fear of death, the boundless life of pleasure and the belief in fate among the people of his time. All the body wants is not to freeze, not to starve, not to thirst. All that the soul wants, is not have fear. Epicurus believes that self-sufficiency is a great good. Luxury is enjoyed most by those who need it least. Epicureanism distinguishes between the following human needs:

- The primary natural and necessary needs, such as food and drink, are always satisfied, as they are essential for peace of mind.
- The non-natural and unnecessary needs (beauty, wealth, power) must always be rejected because they are the source of emotional restlessness.
- The moderate needs between these extremes (eating well, dressing well) shall be satisfied to the extent that they are not too burdensome or too costly. Under no circumstances must man become a slave to his own desires, urges or feelings not even the ethically positive ones, like love or generosity.

The teachings of Epicurus consistently eliminate mythical thinking, and are still regarded today as patterns for naturalistic thinking.

1.7.4   Stoa

The school of philosophy of the Stoa, founded by Zenon of Kition (335–262 B.C.), was named after the portico in the center of Athens (Fig. 1.18), where the teachers taught.

The Stoa is one of the most important Hellenistic schools and was formed in three periods:

- In the ancient Stoa, Cletuses of Assos (330–230 B.C.) and Chrysipp (279–204 B.C.) systematized the teachings of Zenon. They drafted a philosophy that is reminiscent of that everything has its purpose. The destiny (Fatum) is the law of the cosmos, according to which everythinghappened, everything is happening and all that is to come will come.
- In the Middle Stoa, the teaching was expanded by Panaitios of Rhodes (185–98 B.C.) and Poseidonios of Apameia (135–50 B.C.) and absorbed elements of epicurean and oriental origin and was transferred to Rome.

Fig. 1.18 Columned hall in Athens

– The late Stoa was particularly influenced during the Roman Empire by the poet Seneca, the slave Epictetus and the Roman Emperor Marcus Aurelius. In the center of the late stoic philosophy are the mastery of life and moral questions.

Stoic morality prescribes to live "according to nature" or that principle of rationality that the Stoics considered fundamental to man and the entire universe. In contrast to Epicureanism, which in bliss had the goal of existence, stoicism distinguishes between.

- *Dutiful behavior,* which should always be aimed for: in family commitments, in their commitment as citizens, towards the fatherland, in agreements and in friendship,
- *Unjust behavior*, i.e. behavior against reason which is always to be avoided, which includes all emotionally determined actions,
- *Indifferent behavior,* which is neither virtuous nor vicious and relates to things which the wise man does not care about (e.g. wealth/poverty, health/illness, beauty/ugliness) The wise man does not strive for money, he accepts with equanimity the fate of his life.

The Stoa saw in the *Logos*—which was already described by Heraclitus as the "all-founding principle"—the rational principle of the ordered cosmos from which as a resting point all activities emerge. The logo stands in contrast to the myth, which is not supported by rational evidence.

As a result of the conquests of Alexander the Great (*356 B.C. in Macedonia, †323 B.C. in Babylon), Greek culture and stoic thinking had developed over the whole so-called civilized world, Fig. 1.19. The early Stoics were mostly Syrians, the later Romans. The philosophy of the Stoa had its peak with the influential writings of

Fig. 1.19 The empire of
Alexander the Great around
320 BC

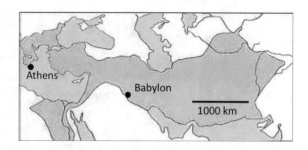

the Roman philosopher and statesman Marcus Tullius Cicero (106–43 B.C.) which
reached the Roman Empire and the Renaissance, and continues to have an effect
until modern rationalism.

1.7.5 Skepticism

The school of thought of skepticism was founded by Pyrrhon (ca. 365–270 B.C.). He
served as a soldier under Alexander the Great, whose campaigns took him to India.
In doing so, he got to know many peoples with different human opinions based
on different ethnic and cultural backgrounds. The skepticism developed by Pyrrhon
was intellectually underpinned by his student Timon of Phleius (320–230 B.C.).
Skepticism as a critical method is directed against the philosophical schools of the
Epicurean and the Stoic. The skeptics, in the face of all the claims that made a "truth
claim", require the application of a skeptical testing procedure with the following
steps:

1. To a given statement with a claim to truth, an opposite statement is found, which
   refers to the same object as the given statement.
2. The admissibility of both statements is established.
3. If it is impossible to decide on one of the two statements, an abstention is made.
4. From the abstention from judgement follows the "peace of mind" of man.

   The ancient philosophical skepticism had a system of arguments (tropics), with
which all possible claims to knowledge from sensual perception, logical conclusions
or the area of cultural customs or traditions were rejected.

1.7.6 Philosophy in the Roman Empire

The Roman Empire, i.e. the territory dominated by the City or the Roman State,
extended over territories on three continents around the Mediterranean Sea, Figs. 1.20
and 1.21.

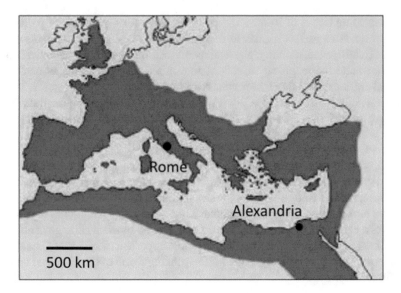

Fig. 1.20 The Roman Empire around 200 A.D

1. Roman kingdom, about 750 to 500 B.C.
2. Roman republic, 509 B.C. till the decline of the republic because of civil war 133 B. C.
3. Roman Iron Age, 27 B.C. till 284.
4. Late antiquity, from 284 to the 7th century, migration (375 to 568), spacing of the Empire into Western and Eastern Roman Empire (395).
5. Downfall of Western Roman Empire (476/480).
6. Transition of the Eastern Roman Empire to the Byzantine Empire (early 7th century: Roman political system + Greek culture + Christianity.
7. Capture of Constantinople by the Osmane (1453).

Julius Caesar, 100 – 44 B.C.

Nero, 37 – 68   Marcus Aurelius
121 – 180

Fig. 1.21 The Roman Empire: Historic epochs and examples of Roman coins

Culture and trade flourished, especially during the imperial period, the "Roman Iron Age". Greek Philosophy was taken up and developed further by the Romans. The poet-philosopher Lucretius (96–55 B.C.) saw philosophizing as a naturalistic world view, as a practical guide to individual life style and as social criticism. In his didactic poem *On the Nature of Things*, which is epicurean ethics in Latin, he contributed to the growing influence of philosophy among Roman poets and statesmen. Marcus Tullius Cicero (106–43 B.C.), Roman philosopher and statesman, took path to the teachings of the Stoics and the Skeptics. He is a representative of the Roman ideal of humanity and propagated a combination of comprehensive spiritual and moral education, general philanthropy, and distinguished manners. Cicero is regarded as the outstanding orator of Roman antiquity. He stood for a synthesis of philosophy and rhetoric. For a perfect speaker he demanded a holistic education, suitable natural abilities (intelligence, mobility, physical merits), knowledge of the theoretical basics of rhetoric and the technical aspects of speech exercises to train the mental and physical abilities necessary for speech are of importance.

Lucius Annaeus Seneca (4 B.C.–65 A.D.), Roman philosopher, poet and statesman and Marcus Aurelius (121–180), Roman emperor, belonged to the Roman social upper class, the Greek Epictetus (35–138 AD) was a freed slave. In their oral or written utterances, they spread stoic ethics. Boethius (480–524) lived in Italy during the Migration Period and practiced under the Ostrogoth king Theoderic a high political office. He was the last major Roman philosopher and underlined the ancient ideal of philosophical wisdom, which—by striving after insight into the causes of the universe and through a morally exemplary life—can show people the safe way to happiness. He professed Christianity, but separated the spiritual heritage of antiquity from Christian faith issues. He became the most important mediator of Greek logic, mathematics and music theory to the Latin-speaking world of the Medieval until the twelfth century.

1.8 Thinking and Believing

The classic antiquity is followed by the epoch of the Middle Ages. This is usually referred to as the time between the end of the Western Roman Empire (476) and the collapse of the Eastern Roman Empire (1453). The invasion of Islam [founded by Mohammed (570–632)] into the Mediterranean world disintegrated the Greco-Roman unity of antiquity. New forms of thinking and believing developed in a geopolitical system with several major powers, Fig. 1.22:

– The Eastern Roman or Byzantine Empire with the capital Constantinople (from 330), Greek language and Christianity as state religion (391). From there follows the Christianization of the Slavic peoples (Serbs, Bulgarians, Russians) with a Greek- Orthodox Church Constitution and the Cyrillic script, modified by the Greek script and named after Cyril of Thessaloniki, (826–869).
– The caliphates and empires of the Muslim Arabs, to whom Mohammed promising the paradise to whom who spread his doctrine. In about hundred years, the Muslims conquered Syria, Palestine, Persia, Mesopotamia, Egypt, North Africa and large parts of Spain.
– The Frankish Empire with the coronation of Charlemagne, King of the Franks and the Langobards, to the new "Roman Emperor" in the year 800, combined with a cultural renewal on the basis of the Latin language, the ancient tradition and of Christianity.

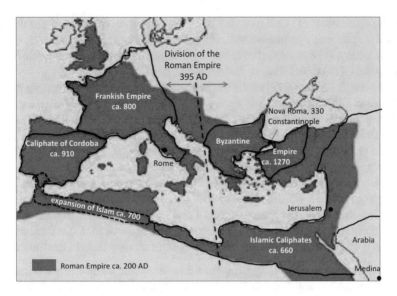

Fig. 1.22 The occidental world from the Antiquity to the Middle Ages

1.8.1 Holy Scriptures

From a cultural point of view, the Middle Ages are a more than 1000-year-old epoch with dominance of faith. With reference to Abraham, the monotheistic religions Judaism, Christianity and Islam are based on "Holy Scriptures", which are regarded as "Words of God" and which are protected by "Canonization", i.e. normative determinations of the contents of Scripture to become the basis for the practice of faith.

- The Koran (Quran) is the youngest of the "Holy Scriptures". It was written in Arabic language in the period from 630–650 and is until today according to the faith of the Muslims the literal revelation of God (Arab. Allah) to the prophet Mohammed. The Koran-based Islam sees itself as a "holistic religion" and demands that the "divine, unchangeable rules of the Koran" and the "Islamic law of the Sharia" rule all aspects of life: religion, politics, economy, law, contact between man and woman, education and upbringing. Muslims have no religious freedom, apostasy from the Islamic faith can be subject to civil law (succession, marriage) and have criminal consequences. Mission for other religions is prohibited to Muslims and may be punishable by death. In the classical Islamic Legal doctrine, the *Jihad,* the fight against the non-Muslim world for defense and expansion of Islamic territory until Islam becomes the dominant religion.
- The Bible originated in the course of about 1200 years in the eastern Mediterranean area and in the Middle East. The Hebrew Bible (Tanach) consists of the three main parts Torah (instruction), Nebiim (book of prophets), Ketuvim (book of psalms) and corresponds to the Old Testament of the Christian Bible with the ten

commandments: (1) God as sole Lord, (2) prohibition of abuse of the name of God, (3) Sabbath commandment, (4) parental oath, (5) ban on murder, (6) ban on adultery, (7) ban on theft, (8) ban on false testimony, (9) prohibition of desire (woman), (10) prohibition of desire (house, goods).

- The New Testament (NT) of the Christian Bible is a collection of scriptures from Original Christianity. It proclaims—with reference to the Old Testament - the Incarnation Jesus Christ (3 B.C.-30 A.D.) as the Son of God and his crucifixion for the redemption of the world from guilt and sin. Jesus preaches meekness, seeking justice, peacefulness, mercy, love of neighbor (Sermon on the Mount). He confirms and deepens the Ten Commandments of the Torah and formulates the Lord's Prayer. The NT consists of the four canonical gospels (Mark ca. 70, Mathew, Luke 75–95, John around 100) about the work of Jesus. It also contains the Book of Acts, the letters of Paul (Pauline doctrine, later important for the Lutheran Reformation) and the apocalyptic Revelation of John. On these as Word of God understood testimonies of the Bible refer all directions of the Christianity.

The faith based on the Holy Scriptures does not value evidence, but is based on "revelatory truths" and "dogmas". Nearly all major religions contain a "doctrine of salvation", according to which—contrary to today's, international law guaranteed human right of religious freedom (UN Declaration)—humankind is divided into two classes:

- "Believers" have salvation in life and, after death, heaven is open to them,
- "Unbelievers" fall prey to damnation (hell), they must therefore be pushed missionary to the "true faith" of the believers.

Missionary religions, such as Christianity or Islam, start from the claim of the "universal truth" claimed by their faith and feel that they are called to "convert" non-believers or people of other faiths into their own religion.

The missionary order of religions can even go against the commandment "Thou shalt not kill" and instigate "religious wars" (e.g. Christian crusades or Islamist suicide bombings) as "religious missions."

- *The self-righteousness of moralists often proves to be abysmally evil* (Carl-Friedrich von Weizsäcker).

1.8.2   Philosophical and Religious Ideas

An attempt to connect the biblical religions with the world of thought of antiquity, Philo of Alexandria (30 B.C.–50 A.D.) undertook a new venture, starting from a Greek translation (Septuagint, third century B.C.) of the Hebrew Bible. He overcomes the difficulty of a synthesis of philosophical (impersonal) and religious (personal) conception of God by postulating that the essence of God cannot explained in the form of rational evidence, but may be directly "perceived". Whoever is not able to see God, should at least try to image "the most holy Logos" whose work is the world. The Logos—already called by Heraclitus a "cosmological, all-reasoning and determining principle"—is occasionally referred to by Philo as the "first-born son of God", he is the bridge from transcendent God to material reality. Also the Hellenistic Judaism refers to "Logos" as the eternal thinking of the one God. Accordingly stands at the beginning of the Gospel of John of the New Testament in Greek also the term "logos", which Martin Luther translated in the Luther Bible (Luther 1545) with "Word of God". The Zurich Bible of 2007 combines the terms as follows: In the beginning was the Word, the Logos, and the Logos was with God, and of God's essence was the Logos.

A complex philosophical-religious movement emerged in the first century under the *Gnosis* (Greek: knowledge) as a mixture of Christian ideas and Greek mysteries together with Persian, esoteric-Jewish and Egyptian influences. The Gnosis believes in a perfect all-encompassing God, but the world and Human beings are the work of an awkward "creative spirit" (demiurge). Truly "real" is the spiritual world, the material world is basically void. The Gnosis raises the claim of a knowledge that goes beyond rational insight from the "true reality." An "agnostic" position, according to which it is not possible to know whether there are gods or not, was also the notion of the Sophists (Protagoras) in the fifth century B.C. Agnostics hold the view that existence or non-existence of a higher instance (God) is either unexplained or basically is not to be clarified.

Around the year 200, when the Roman state power disintegrated, the Roman Church formed their organized communities—kind of a state within the State. The leaders of their communities won—as "followers of the apostles of Christ" and Roman-Christian "Fathers of the Church" (Patristic)—spiritual and political authority. From the year 345 Christianity became the sole State religion of the Roman Empire. Already in 325 Emperor Constantine called the Council of Nicaea (today Iznik, Turkey), where the "Trinity of God: Father, Son and Holy Spirit" was claimed, and the "Christian confession of faith" has been made binding on the faithful. In the year 367, the canonization of the New Testament (written in Greek) was defined as "normative saints Scripture" and standard of religious practice. Further fundamental religious-philosophical developments in the Middle Ages are the philosophy and theology of Patristics as well as Scholasticism, which were influenced by Augustine as a school for the synthesis of faith and reason, fundamentally formulated by Thomas Aquinas, Fig. 1.23.

1.8.3 Patristics

Patristics means the time of the "Church Fathers" who were active from the second century onwards. Of these spiritual-religious teachers of Christianity the "Apostolic Fathers" are regarded as immediate disciples of the apostles of Jesus. They were followed by the "apologists" who tried to show that some approaches of ancient philosophy—like Plato's dualism (interpreted as body-soul dualism), the metaphysics of Aristotle and the Stoics' faith in providence—are in harmony with

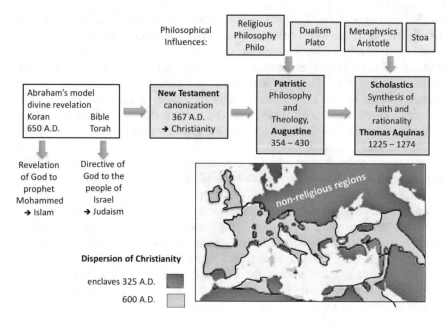

Fig. 1.23 Religious and philosophical developments in late antiquity and in Middle Ages

Christian doctrine. One even goes to such lengths as to accept the Delphic motto already used by Socrates oracles "Know thyself" as a divine revelation. With the dogmas of the councils in the fourth century, the systematization of the Christian teachings takes place and the efforts of the church apparatus to "direct the souls" of believing Christians.

The systematic processing of the Neo-platonic heritage in the context of the Christian Faith teachings were given by Augustine of Hioop (354–430), the most important Latin Father of the Church. With his work "About the God state" he created the first text which shows history—in a departure from the ancient cyclical conception of time and the principle of eternal return—as a sequence of unique and unrepeatable events.

According to Augustine, the history of mankind is marked by the battle of two "virtual Empires" in which man can live according to his own decision:

- The earthly state is the society of the devil. It corresponds to nature, to matter, the bodies of individuals, factual history and the economy.
- The heavenly state is the state of God, the society of the just, the eternal, the divine revelation.

The state of mankind finds its only reason for existence in the development of the state of God and to promote the Christianization of humanity according to the plan to support providence. According to Augustine, the following theorems are of particular importance:

- The equalization of Plato's "ideas" with God's thoughts and the rejection in principle of skeptical relativism,
- The assertion of the independence of the "spiritually recognizable world" vis-à-vis the sensually graspable,
- The interpretation of cognition as the processing of divine inspirations and enlightenments (Theory of Illumination),
- The mutual support of faith and knowledge: "see to believe, believe to see",
- The systematic application of the Trinity principle to considerations of God, the world and human consciousness,
- The radical devaluation of physical pleasure and the corporeality of the human being towards the soul-spiritual,
- The foundation of human happiness in the love for and joy of God in an afterlife perspective,
- The interpretation of the historical process as linear progress on the conflict-laden way to eternal salvation.

Augustine's reflections and theorems are today known under the term *Augustinism*. The general aspects of Augustinism are the dualistic division of reality and the principle derived from it of transcending the sensual world towards the invisible world.

Between the 5th and the sixteenth century, in the geographical area of Europe and West Asia's new dominant forms of religious and cultural life were developed. The main languages of philosophy at that time were Arabic, Latin and Hebrew. Between the 9th and the twelfth century, Arab and Jewish philosopher's combined thoughts of ancient Greek-roman philosophy with own innovations. This was achieved in the twelfth and thirteenth century by translations of Arabic and Greek texts in the Christian-Latin world. In Bologna, at the end of the twelfth century, the first university with the teaching methods *lecture—exercise—disputation* was established. In the thirteenth and fourteenth centuries with scholasticism, philosophy received new forms, contents and methods, which were particularly developed at the universities of Paris and Oxford.

1.8.4 Scholasticism

Scholasticism is a school of philosophy in Europe from about 1100 to 1700 aiming at harmonization of Christian theology with classical and late antiquity philosophy, especially that of Aristotle and Neoplatonism. It is a dialectical procedure with the following steps:

1. Formulation of the question,
2. Providing arguments for an affirmative answer (Pro-arguments),
3. Giving arguments for a negative answer (Con-arguments),
4. Formulation of a problem solution with corresponding arguments,

5. Resolving arguments (e.g. illogical or conceptually unclear) which are not
   consistent with the solution.

"Scholastics" are originally the teachers at the convent schools who train priests.
To distinguish the teaching taught in monastic religious communities from the university theology, the term "scholastic theology" was coined. It wants to show the application of philosophy to the "revelatory truths of the Holy Scriptures" to gain insight
into the contents of faith and bring them closer to the thinking human spirit.

The "supernatural truth" is from the outset fixed in Christian dogma. To do well
is to obey God and this highest good is also the highest value. The Scholasticism
does not want to find new truths, but to systematically describe "supernatural truth"
and disprove objections on the part of reasons against the content of the revelation.
In scholasticism, God is the epitome of perfection, but he can be examined with the
human intellect. (In contrast, "negative theology" claims the incomparability of man
and deity. Since God is absolute, he can't be defined).

Thomas Aquinas (1225–1274) is considered the most important scholastic. He
was a Dominican monk and claimed to give theology the rank of a science. The
philosophical–theological teaching direction (Thomism) founded by him replaces
the Aristotelian pair of terms *potential* (possibility, capability) and *actuality* (reality)
by the pair of *essence* (the character of a thing) and *existence* (reality of a thing). The
metaphysics of Thomism proceeds from the diversity of the obvious being (stones,
plants, animals, humans). According to Thomas of Aquinas, a human being is the
substantial connection of body (with Aristotle: matter) and soul (with Aristotle:
form). Through his body, a human takes part in the world of matter and through his
soul in the world of pure spirit (body/soul dualism).

Thomas Aquinas postulates that faith and reason are not contradictory, since both
are from God. Therefore, theology and philosophy cannot become different truths,
but they differ in the method: philosophy starts from created things and thus reaches
God, theology begins with God.

The life cycle of every being—in Aristotelian terms the transition from potential
to actuality - is marked as the transition from *essence* to *existence*. This means with
regard to the universe that there is a primordial being (God) that creates without to
have been created in his part. God unites as the only "perfect being" essence and
existence. Since God would not be perfect without existence, he must exist. This is the
so-called "ontological proof of God". In his ethics, Thomas Aquinas distinguished
between theological and natural virtues.

- The cardinal theological virtues—faith, hope, love—are most important and only
  accessible under the grace of God, where love is the key to all human activities
  to a final divine purpose.
- The natural cardinal virtues represent the best possible expressions of the human
  wealth. Reason, then, is assigned to wisdom and knowledge, and justice and
  moderation is assigned to desire. The virtues determine the inner attitude of a
  human; the outer order and the human actions are guided by laws.
- The supreme principle of practical reason is to do well and avoid evil.

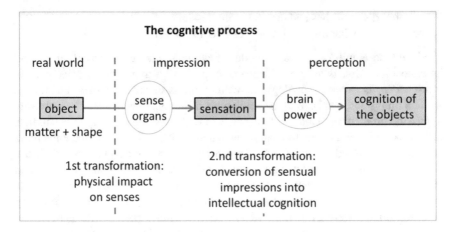

Fig. 1.24 The cognitive process according to Thomas Aquinas

1.8.5   The Cognitive Process

According to Thomas Aquinas, the process of knowledge can be represented as follows: An object of the real world is first perceived by a sense organ. From the sense organ it reaches the general sentience and is held as a single image in the imaginativeness. The intellect now abstracts the general form from the single image and thus enables the cognition of the thing. The process of cognition thus runs through two "transformations", which Thomas described as 1st difference and 2nd difference in the process of cognition, Fig. 1.24:

- In the first transformation, the physical-empirical action produces an effect on the fundamentally different sensory level.
- In the 2nd transformation, the sensual perception is converted into intellectual-spiritual knowledge.

For the description of the process of knowledge, Wilhelm von Ockham (1285–1347) developed the "principle of ontological thrift". In attempts of explanation, only as many assumptions should be made as are absolutely necessary: "You should not postulate more entities than necessary".

1.8.6   Infinity of the Universe and Pantheism

A departure from strict scholasticism is represented by Nicolaus von Kues, Latinized Cusanus (1401–1465). He developed the theory of the "rational unrecognizability of God". It proceeds from the stages of human insight:

– sensual perception initially gives unrelated impressions,

- the mind distinguishes and orders the impressions,
- reason unites what the mind separates into a synthesis.

Since God is infinite, he can simultaneously combine every possible thing and its opposition. God is as "coincidence of opposites" a transcendent principle. As *transcendent* is called in theology, something that is outside the world and all the world's things and lies beyond the limitations of human experience. The only way to face the problem of "God" with reason is to make assumptions. By this, Cusanus means analogies of a geometric nature, which make it possible to compare the finite and the infinite (God) speculatively. For example, if one models God by an ideally circle, then the human mind and the things of the world are alike a polygon inscribed in the circle. However much you increase the number of sides of the polygon, the polygon construct of the human mind will never attain the ideal circle representing God. A similar modeling can be done by comparing a finite triangle with of an infinite straight line, Fig. 1.25. These models suggest that in God all things unite.

A further development of the model of the "coincidence of opposites in infinity" was made by Giordano Bruno (1548–1600). He abolished the Aristotelian contrast of the (sublunar) earth to the (supralunar) heaven and breaks through the "seclusion of the ancient and the Christian world view" by giving the world (the universe) infinity. He developed the following theses:

- The universe has no outer limits. There is no barrier that encloses itself.
- The cosmos is infinite, and the countless stars spread out in space into all directions. Space thus is acentric and has no privileged point.
- The universe is homogeneous in each of its parts. The stars with their planetary systems are dispersed in a vacuum and not enclosed in crystalline spheres, as Ptolemaic astronomy had claimed

Bruno represented a pantheistic world view: the infinite universe is God. He is eternal and unchangeable. The changeable individual things have only a temporary part in this. With these theses Bruno contradicted the doctrine of the institution Church with its dogmas of "Creation", the "Hereafter" and the "Last Judgment".

Because of the pantheism he proclaimed, Giordano Bruno was condemned by the pontifical Inquisition (guilty of heresy and magic) and was burned at the stake in Rome in 1600. After 400 years, in the year 2000 this was declared an injustice by the papal cultural council. Today, Bruno is regarded as a pioneer of the world view of modern times.

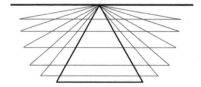

If the sides of a triangle are extended to infinity, the triangle is completely like a straight line. In a geometric-philosophical analogy, Cusanus speculated that God is symbolically both triangle an straight line because all separate and different things coincide if extended to infinity.

Fig. 1.25   Model by Nicolaus von Kues of the coincidence of opposites in infinity

1.9   The Rise of Technology

In addition to concepts of philosophy and physics embedded in antiquity, important root of technology evolved in the age of Renaissance. The Renaissance was a period in European history marking the transition from the Middle Ages to Modernity and covering the fifteenth and sixteenth centuries.

The natural philosophers of that time identified "machines" that can facilitate life and workload of humans. Also in these times, the term "engineer" appeared, coming from Latin word 'ingenium' that means a "stroke of mind leading to an original idea". A machine (or mechanical device) is a mechanical structure that uses forces and control movement to perform an intended action. The classical simple machines of ancient technology are *Lever, Wheel and Axle, Pulley, Inclined plane, Wedge, and Screw.*

1.9.1   Machine Elements of Leonardo Da Vinci

Leonardo da Vinci (1452–1519) is revered as outstanding polymath of all times. He wrote the first systematic explanations of how machines work and how the elements of machines can be combined. His tremendous talents as an illustrator allowed him to draw his mechanical ideas with exceptional clarity. Leonardo described and sketched ideas for many technology inventions ahead of their time.

Four hundred year after Leonardo, Franz Reuleaux (1829–1905) is recognized as the precursor of modern engineering design with his books on *Theoretical Kinematics and Fundamentals of Machinery.* As the first scientist, he based on mathematical and scientific fundamentals, tried to link up the generating of ideas, kinematics of machines, and engineering design into one coherent whole. Reuleaux put in order the basic elements of all machine in 22 fundamental classes. They are named in Fig. 1.26 together with drawings made 1492 by Leonardo da Vinci and rediscovered in1965 (Codex Madrid I). It can be seen that Leonardo da Vinci made drawings of almost of all machine elements long before they were applied in modern technology.

However, only a few of Leonardo's brilliant designs could be realized during his lifetime, as the modern scientific approaches to metallurgy and engineering were only in their infancy during the Renaissance. The engineering realization of high performance machine elements as we know them today became only possible after the development of steel (e. g. Bessemer steel 1895), that can be hardened to withstand the high working loads and stresses in machines.

Today, many machine elements are designed in line with Leonardo' s ingenious technology design drawings. Figure 1.27 shows two important examples:

– Leonardo da Vinci invented the ball bearing between the years 1498–1500. He
 designed it to lower friction in rotary machine elements. Today, ball bearings are

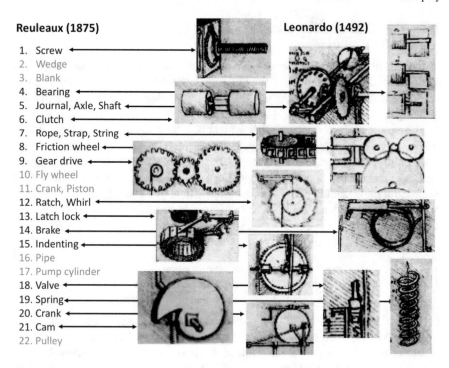

Reuleaux (1875)   **Leonardo (1492)**

1. Screw
2. Wedge
3. Blank
4. Bearing
5. Journal, Axle, Shaft
6. Clutch
7. Rope, Strap, String
8. Friction wheel
9. Gear drive
10. Fly wheel
11. Crank, Piston
12. Ratch, Whirl
13. Latch lock
14. Brake
15. Indenting
16. Pipe
17. Pump cylinder
18. Valve
19. Spring
20. Crank
21. Cam
22. Pulley

Fig. 1.26 The basic machine element and drawings of "Elementi macchinali" by Leonardo da Vinci

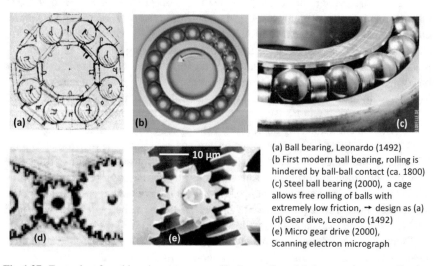

(a) Ball bearing, Leonardo (1492)
(b First modern ball bearing, rolling is hindered by ball-ball contact (ca. 1800)
(c) Steel ball bearing (2000), a cage allows free rolling of balls with extremely low friction, → design as (a)
(d) Gear dive, Leonardo (1492)
(e) Micro gear drive (2000), Scanning electron micrograph

Fig. 1.27 Examples of machine elements proposed by Leonardo and their contemporary realization

indispensable to realize rotary motion in many technology areas, from automobiles, trains and wind energy plants to fans in air conditioning of buildings and computers.
- Gear drives are needed to convert force, torque and mechanical power in large scale machinery as well as in mechanical microsystems.

All these engineering applications utilize technology principle first proposed by Leonardo da Vinci.

1.10 The Turn to Modern Times

The modern age begins with the discovery of the real heliocentric world view by Copernicus (1543), Fig. 1.28. The new world outlook, which is derived from the heliocentric world view, could not be brought into harmony with the ancient and medieval image of the place of man in the cosmos. The model of the cosmos with the sublunary sphere for the human beings and the superlunary sphere for God and the angels broke. "For the first time in history, man on this earth faces only himself" says the physicist Heisenberg (1901–1976) and the American philosopher Thomas Kuhn (1922–1996) calls the Copernican revolution a "paradigm shift of world history".

The Copernican revolution can be seen as the beginning of the—several centuries-enduring—transition from the ancient cosmos to the "open world view" of modern times. In addition to the revolution in the macrocosm through the heliocentric world view, with the invention of the microscope (ca. 1595), the "microcosm" also was opened to modern thinkers. Pascal (1623–1662) formulated as a simple thought experiment the idea of the world's intertwining to infinity. Leibniz (1646–1716) spoke of an "imperceptible world", Berkeley (1685–1753) of a "new world", and according to Hume (1711–1776), microscopes open up a "new universe in miniature". Against the background of all these developments, Kant (1724–1804) presented the elementary new question in the world of **PHILOSOPHY:** What can I know?

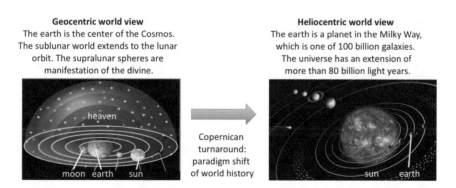

Geocentric world view
The earth is the center of the Cosmos. The sublunar world extends to the lunar orbit. The supralunar spheres are manifestation of the divine.

heaven

moon   earth   sun

Copernican turnaround: paradigm shift of world history

Heliocentric world view
The earth is a planet in the Milky Way, which is one of 100 billion galaxies. The universe has an extension of more than 80 billion light years.

sun   earth

Fig. 1.28 The geocentric and heliocentric world view in simplified overviews

The geographical knowledge of the earth expanded. The spherical shape of the Earth was confirmed by a circumnavigation of the globe (Magellan 1522) and the portable chronometer was invented for the measurement of time (Henlein 1512). With Galilei (1564–1642), the scientific research of the laws of nature started. He founded the theory of motion (kinematics) and the theory of elasticity, and defends the heliocentric system in a dispute with the pope. The law of gravity—which explains why the planets move around the sun and why on earth all things fall down—was discovered by Newton in 1666. From the multitude of physical observations and verifiable measurements, the world of **PHYSICS** evolved.

Also in the areas of artistic and technical skills, which were called *techne* in antiquity, there were new developments in the Renaissance. In painting, sculpture, and architecture, with the application of perspective representation (Brunelleschi 1410) completely new works were created. The letterpress with movable letters (Gutenberg Bible 1454) opened up new possibilities for literature and compositions. First engineering works and models (e.g. gears, ball bearings, clockworks, bicycle, flying machines) of the ingenious polymath Leonardo da Vinci (1452–1519), who was also painter (Mona Lisa), architect, anatomist and natural philosopher, were harbingers of the emergence of the world of **TECHNOLOGY**.

Chapter 2
Philosophy

Philosophy attempts to understand what we think and what we do. The world of philosophy is characterized by the relationship man–nature–idea. The chapter presents the core topics of philosophy: philosophy of being, philosophy of the self and philosophy of mind.

- "Philosophy begins when human beings start trying to understand the world, not through religion or by accepting authority but through the use of mind and reason". (Bryan Magee in his book on the Story of Philosophy).
- "Philosophy is not reflection on an isolated thought, but on the total of our thoughts. Each of the great philosophers has described this whole in his own way "(Carl Friedrich v. Weizsäcker in his book The Human being in his History).
- "The philosophical models of the past arose from the knowledge of that time and therefore corresponded to the way of thinking to which such knowledge has led. Therefore, the terms to which the philosophers have been led by an analysis of their experience of nature, could not foresee the phenomena which can only be adapted with the technical aids of today" (Werner Heisenberg in his book Physics and Philosophy).
- "Within philosophy one could never agree on a single method" … states the textbook Philosophy—History, Disciplines, and Competencies of 2011 (ed. Breitenstein and Rohbeck) … "Philosophy has been always driven by similar questions, since the beginning of the Greek philosophy of nature which it could never fully, satisfactorily answer. The philosophers argued depending on the problem and temperament *phenomenological, transcendental, dialectical, analytical*, etc. Thus, on the one hand there are hardly any problems, which simply were solved and filed and on the other hand there is no linear progress".
- "Philosophers are engineers of ideas" … says the American philosopher Simon Blackburn in his book Thinking—The Big Questions of Philosophy, … "for just as an engineer studies the structures of material objects, so does a philosopher study the structures of thought. Philosophers try to explore the structures, which are forming our world view. Our ideas form the spiritual home in which we live".

© The Author(s), under exclusive license to Springer Nature Switzerland AG 2021
H. Czichos, *The World is Triangular*,
https://doi.org/10.1007/978-3-030-64210-5_2

2.1 Dimensions of Philosophy

According to the British philosopher and logician, Alfred N. Whitehead (1861–1947), modern philosophy exists essentially "from a series of footnotes on Plato". Arno Anzenbacher presents this in his Introduction to Philosophy (2010) graphically by the Platonic Triangle, Fig. 2.1.

Starting from the three corners of the Platonic Triangle, the main directions of theoretical philosophy can be outlined in a simplified way as follows:

- Philosophy of being: Here, thinking about the world asks for the "existence" that is based on observable phenomena. This is the approach of classical *Metaphysics* founded by Aristotle, which today is called *ontology* (the theory of being).
- Philosophy of the self: This direction of philosophical thinking begins with the "ego (the I)"—in the language of philosophy also known as "subject". The main models are Rationalism (Descartes, Leibniz, Spinoza) and Empiricism (Locke, Hume, Berkeley). The connection between rationalism and empirical research undertook Kant in the time of the classical German idealism period of "Enlightenment" with his epistemology. A special variant of the ego philosophy is existentialism (Heidegger, Sartre).
- Philosophy of mind: Philosophizing here starts from the "idea" and develops philosophical models of the "Absolute" in a synopsis of "Being and Ego" (object and subject). These include Hegel's complex philosophical system, the historical Materialism (Marx), Analytical Philosophy (Russel, Wittgenstein) and the 3-World Theory (Popper).

Philosophical thinking is also used today as "mental reflection", as can be seen from the following word formations: cultural philosophy, political philosophy, philosophy of history, philosophy of law, social philosophy, etc. This "applied philosophy" is not considered here. Neither is there any philosophical reflection on physics and technology. In this second part of the book, the philosophical core topics of the Platonic Triangle are considered.

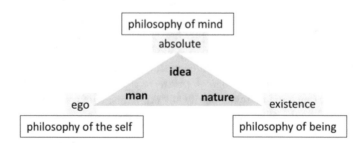

Fig. 2.1 The Platonic Triangle symbolizes with the corners *man—nature—idea* the space of theoretical philosophy

2.2 Philosophy of Being

The doctrine of principles developed by Aristotle (384–322 B.C). can be described as the theory of being. In contrast to the Platonic dualism of the doctrine of ideas (which Aristotle considered "superfluous"), for Aristotle there is no super world that would be of higher value than physical reality: what lies beyond all possibilities of experience can mean nothing for us. We cannot verifiably refer to it or talk about it. There is only one world, namely the world in which we live, and it consists of that very matter which, according to Plato, does not deserve to be examined in the light of ideas. In contrast to Plato's dualism, Aristotle advocates a "horizontal thinking", Fig. 2.2.

2.2.1 Terms

Terms are, according to Aristotle, an important means to distinguish the general from individual things. For example, there are very many different individual houses. The general (universal) term *house* means the totality of all houses and thus characterizes the general, which is common to all concrete individual houses.

Ancient philosophy raised the question of "universality", which has not yet been conclusively clarified, whether there are general terms or whether they are thought constructions, Fig. 2.3.

According to Aristotle, "General" only exists if "Individual Things" also exist. The General arises when a general view of similarity is formed from many thoughts gained through experience. The General is thus an abstraction, contained in the individual things: "Universals are in things". Plato, on the other hand, teaches that there are Universals (Ideas) before individual things—they could be called a "master plan" according to current usage. He illustrates this assumption with his allegory

In his painting *The School of* Athens, Raphael represents with characteristic gestures of Plato and Aristotle the directions of their philosophical schools.

Plato points upwards to the idealistic good and indicated a vertical direction of thought.

Aristotle points with his stretched hand to the world and indicates a horizontal direction of thought.

Fig. 2.2 Symbolization of the philosophies of Plato and Aristotle in a painting by Raphael

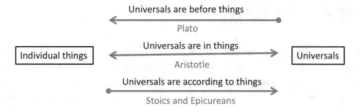

Fig. 2.3 The concepts of singular things and universals in different views of philosophy in ancient times

of the cave, according to which imprisoned cave dwellers perceive only "shadowy figures" as images of the universals. Accordingly, the Universals (the Platonic Ideas) are "more real" than concrete things. Plato postulates: "Universals are before things". Stoics and Epicureans, on the other hand, assume that the general generates in thought will. This point of view can be explained by the formula "Universals are according to things".

2.2.2 Categories

Highest generic term under which all other terms can be subsumed, Aristotle calls "categories." These are elementary expressions that denote something and can therefore make a statement. Besides *having, doing, suffering*, Aristotle calls the following categories:

- Substance, describes the object under consideration ("What it is")
- Quality, indicates what kind of something is what
- Quantity, indicates how large, long, wide, etc. something is
- Relation, interpretable as "interaction" with other objects
- Where: Location
- When: Time determination
- Status: Determine the situation.

2.2.3 Logic

Terms are linked to judgments. In these, a term (subject) is used to describe something specific testified (predicate). Judgments in turn are connected to conclusions. From certain conditions (premises) something new is derived. This "art of Aristotle's conclusion" and of proof is called syllogistics.

Aristotle's syllogism establishes the logic, the science that can show when and why a conclusion is valid or false. A conclusion emerges, when a thought passes from one judgment to the next, with the associated sentences are in a necessary context, so

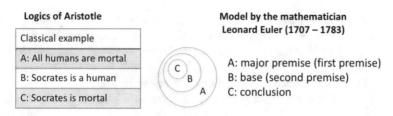

Fig. 2.4 Logic: Classical example after Aristotle and modelling by Leonard Euler

that the conclusion follows from the premises. From this consistency, according to which a preceding sentence is the reason for the conclusions, the logical connection exists.

A syllogism must therefore consist of at least three propositions Fig. 2.4:

- Major premise (first premise, A),
- base (second premise, B),
- Conclusion (C).

The most famous example proves the mortality of Socrates as follows:

- All humans are mortal (A),
- Socrates is a human being (B),
- So, Socrates is mortal (C).

A chain of conclusions, Aristotle calls a proof. This method is deductive, i.e. it goes from the general to the specific. The counterpart is induction; it is the progress from the individual to the general. The induction searches for the common within a genus. The classification of all being enables the definition. It consists of the genus and the species-forming differences (e.g. human is a reasonable living being). Aristotle formulates the principle of contradiction: it is impossible, that the same thing is due and not due in the same respect.

If the logic illustrated by the classical Socrates example is generally applied, the "truth of the premises" must be proven or it must be based on an axiom. The axiom concept, going back to Aristotle, means an "immediately obvious principle", which is "evident" and no proof is needed. In modern axiomatic, axioms differ from other statements only by the fact that they are not derived. Some of these axioms are valid for more than one science. For example, the idea "when A equals B and B equals C, then C equals A" (equivalence principle) is not only valid for numbers, but also applicable to any object.

2.2.4 Rhetoric

Rhetoric (Greek oratory) is a summarizing term for the theory and practice of speech formation and it is important for truth-finding and education of beliefs. Aristotle is convinced that between truth and error, there is a "gap between probability and uncertainty". There are problems, e.g. political and judicial, which by their nature are not "definitive" but only know "temporary" and "probable" solutions. Here, the rhetoric must as a "technique of communication" provide for clarity of presentation and avoid disputes which arise from mutual incomprehension.

The foundation of the entire rhetorical tradition is the thought of Aristotle that a good, irresistible speech must be based on the views of the interlocutor. The rhetoric of the speaker can thus inform the interlocutor of the conclusions of the speech, since these ultimately arise from one's own convictions. The methodology of rhetoric is divided into the following operational steps:

1. Topic, i.e. the search for generally accepted beliefs, Latin *loci communes* (commonplaces). Like rhetoricians in the past, today opinion pollsters, marketing institutes or even the media are looking for suitable topics in order to be able to align political or advertising messages accordingly.
2. Structure: introduction—factual report—presentation of evidence—conclusion.
3. Style that suits the speaker, the audience and the topic and makes the speech new, unusual and interesting.
4. Memorizing the content and form of speech (mnemonics, memory images).

Tips for a good speaker in modern times: *Main clauses, main clauses, main clauses. Clear disposition in your head—as little as possible on the paper. Facts or appeal to feeling, slingshot or harp* (Kurt Tucholsky 1930).

2.2.5 Space and Time

In the conceptual categories of Aristotle, *space* and *time*—articulated as "where" (localization) and "when" (dating)—have a special meaning.

Experience shows that **space** is a fundamental model of order for all physical processes; it is a kind of "container" for matter. The location of an object in space is defined with four basic assumptions:

- The location of a viewed object in space encloses the object.
- The location of each object is neither larger nor smaller than the object itself.
- Every location can be left by the object under consideration.
- Each location is, or is determinedm by the three dimensions: length, width and depth.

From these premises Aristotle concludes that every body occupies a place and also every place must always be occupied by a body (Where there is a body, there can be no other one).

Today, for the clear identification of the location of an object within space, location coordinates in a Coordinate System are utilized, Fig. 2.5. The most commonly used Coordinate System is the Cartesian Coordinate system with rectangular (orthogonal)

Space and location according to Aristotle
Space
Object at a location
in space
Cartesian Coordinate System
z-axis
z
object point with
coordinates x, y, z
y
y-axis
x
x-axis

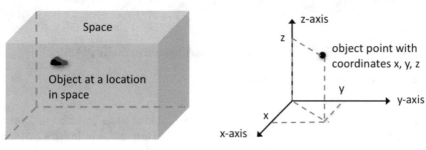

Fig. 2.5 Space and location in the classical view and in the current coordinate system representation

spatial axes, named after Rene Descartes, (1596 -1650) The location of a point P in space is exactly marked with an indication of the coordinate triplet (x, y, z).

Time was historically understood as a concept independent of space. It describes the sequence of events that are subjectively linked to our experience and are perceived as past, present and future. The time view of antiquity is connected with a juxtaposition of (a) being as the permanent and (b) becoming as the changing in coming into being and passing away.

- Plato understands time as the "completion of eternity" in a circular movement, arithmetically describable as a numerically advancing image of the permanent eternity.
- Aristotle also sees time as a number of motions, which is expressed in variable intervals and whose limits are the "now-moments". The now is the transition from the future to the past. The sequence of now-moments forms a continuum that is in principle infinitely divisible.

The idea that time has a "cyclical structure"—such as the periodic return of the seasons or the periodic movements of the heavens—remained a general idea throughout antiquity. It is also expressed in the cyclical division of time into the days of the week, Fig. 2.6.

In the ancient world view, the periodic return of natural processes is an inner necessity for everything that happens—just like autumn in summer follows. This natural-philosophical observation of the vegetation cycles is probably followed by

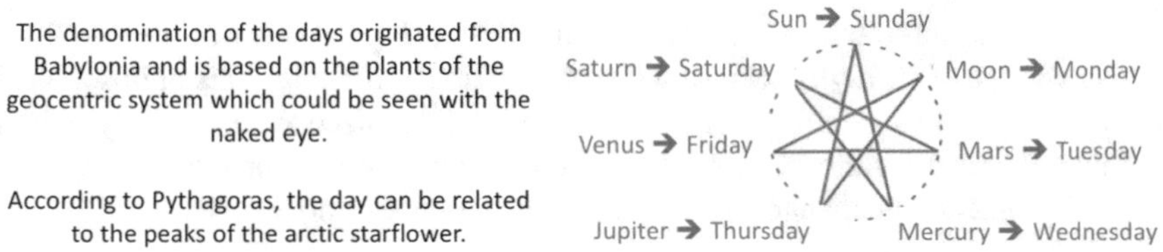

The denomination of the days originated from Babylonia and is based on the plants of the geocentric system which could be seen with the naked eye.
According to Pythagoras, the day can be related to the peaks of the arctic starflower.
Sun ➔ Sunday
Saturn ➔ Saturday
Moon ➔ Monday
Venus ➔ Friday
Mars ➔ Tuesday
Jupiter ➔ Thursday
Mercury ➔ Wednesday

Fig. 2.6 The cyclical division of days in a week

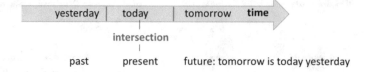

Fig. 2.7 Linear concept of time and points of time

also the mythical-religious belief of the "transmigration of souls", which is a "cyclic reincarnation of the soul". The linear concept of time originated in Judeo-Christian culture and is based on the conviction that there is only one direction of time, Fig. 2.7.

This concept of time also includes the concept of history as an unrepeatable epoch of temporal sequences of events. A linearly progressing time also suggests the concept of "progress", which can be associated with future-oriented thinking and acting. Time thus marks the progression of the present coming from the past to the future: In physics, time, like space, is used to describe physical events: "*The past is factual; the future is possible*" (Carl-Friedrich von Weizsäcker).

Among all conceivable events in three-dimensional space—in combination with all conceivable time sequences—only such events can be observed that obey the laws of physics. Ever since Einstein's Relativity theory, space and time are theoretically described in a unified four-dimensional space–time structure modeled with spatial and temporal coordinates, see Sect. 3.2.5.

2.3 Philosophy of the Self

The philosophy in the seventeenth and eighteenth century is based on different models of thought in two ways from the ego in the platonic triangle, depicted in Fig. 2.1. On the one hand, the "I" is understood as a "rational being" and developed the model of *Rationalism.* On the other hand, one sees the ego as a "sensory being" and thus establishes *Empiricism.*

2.3.1 Rationalism

Rationalism assumes that—under the condition of a "logical order of the world"— the knowledge of reality is achievable through rational thinking. This philosophical model assumes that only then can something be recognized with certainty if it is possible to grasp it rationally and to move on from simple, immediate insightful principles that are independent of or precede experience (a priori).

2.3.1.1 Descartes

The founder of rationalism, Rene Descartes, Latinized Cartesius (1596–1650), sees as the basis of true knowledge the own self: "I think, therefore I am" (cogito ergo sum). While the ancient philosophy of nature attempted to create a "uniform basic principle" to bring order to the variety of things and phenomena, Descartes represents a "dualism" of body and mind. The ego is the "recognizing subject" in which spirit, soul, mind, reason go together. The counterparts are the "objects" of the outer Nature. He interprets the human body as a "mechanism of nature," separates it from the mind and postulates two different substances: (a) the "extended substance", which forms the physical part of the human body and (b) the "thinking substance" which comprises God and the human mind.

- The body is under the influence of natural laws.
- The mind is free and makes its decisions through reason.

This division of nature into two parts—spirit and matter, subject and object, observer and observed—became an essential part of the modern occidental world view. The question of mediation and interaction between these substances remains open. This is the so-called the *body-mind problem.*

In addition to his philosophical work, Descartes founded the analytical geometry and designs the method of Cartesian analysis and synthesis of problems:

- Decomposition of a problem into its simplest elements (reductionism),
- Analysis of the individual elements to obtain clear and simple statements,
- Composition of the element statements for the overall solution of the problem.

   (It must be noted that the application of the classical analytical procedure of reductionism depends on the condition that "interactions " between the elements are non-existent or, at least, weak enough to be neglected).

2.3.1.2 Spinoza

In contrast to Cartesian dualism, the philosophy of Spinoza (1632–1677) combines both God and matter, and thinking and matter to the model of a "monism" in which God and the world form a single "substance". The substance contains internal differentiations: the "expansion", whose basic modes are "shape and movement", and the "thinking" with the basic modes "idea and act of will".

According to Spinoza, it is irrelevant whether the world is created with the help of spiritual, abstract, religious concepts or described by means of the terms for material objects. God is neither outside nor inside the world: he is the world and directs it by the laws of nature. All events are subject to an absolute logical necessity. As creatures of god we embody the same duality. We are bodies in one and the same person and soul. It is as if the body is the soul in another form. They are just different description types of the same reality.

In his epistemology, Spinoza distinguishes three types of cognition:

(a) Sensual cognition, which is achieved through affections (effects on sensation) and can give rise to disordered concepts.
(b) Rational knowledge operating with defined common concepts.
(c) The intuitive cognition that gains cognition in relation to the Absolute.

According to Spinoza, only true ideas are clear. They conclude the certainty of truth, since truth is its own standard and has no other criterion except him.

2.3.1.3 Leibniz

Rationalist philosophy reached its peak with the polymath Leibniz (1646–1716). In mathematics, he invented (parallel to Newton) the differential and integral calculus, and became one of the most prolific inventors in the field of mechanical calculators. He also projected the binary number system, which is the foundation of all digital computers and last the "digital world", as illustrated in a simplified picture in Fig. 2.8.

Leibniz postulated in his philosophical considerations, the "monad theory". In doing so he starts from atomism, which for the explanation of physical existence uses atoms as simple last units. Leibniz extends this postulate and argues that material atoms are real but no "points", points are indivisible and mathematically unreal. Reality and indivisibility are united only in elementary substance units, the *monads*. They are individual "force points", have no form, can be neither generated nor destroyed and are "windowless", nothing can get out of them or into them. Organisms are complexes of monads. They have a central monad (soul), which, together with the other monads make up the "body", in one of God "pre-stabilized harmony".

As a contribution to epistemology, Leibniz adds the empirical formula "nothing is in the mind that has not previously been in the senses" by the addition "except the mind itself". The mere stringing together of sensory impressions only leads to probable results (called factual truths), clear and correct results (called truths of reason) only reason can recognize. The truths of reason—to which the logical laws count—are necessary, but the factual truths are not without a comparison with reality.

Leibniz found that there are two principles for logical reasoning: (a) the principle of consistency (a sentence is either true or false), (b) the principle of sufficient reason

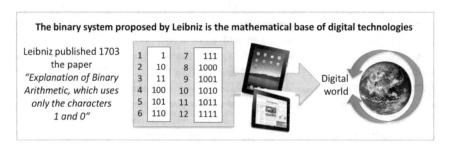

Fig. 2.8 The binary numerical system and its follow-up

(without sufficient reason no fact can be true and no statement can be right). The last sufficient reason must be God. Behind this is the conviction that God has chosen and created the "best world" of the logically possible worlds. Its basic principle is that nothing happens in it without sufficient reason. However, this raises the problem of theodicy (justification of God), i.e. the question of how God's goodness is compatible with evil in the world. Leibniz sees the solution in the fact that the "best of all possible worlds" is by no means a world which can only contain perfection, because then it would consist in a doubling God. So if a world is created by God at all, then this is only possible under admission of the evil is possible.

2.3.2 Empiricism

In empiricism, which developed historically in England, the sensual experience is considered as the basis of the cognitive faculty of humans. Empiricism assumes that reality can only be perceived with the help of individual objects and sensory perception. Thinking has only the task to take the impressions to order correctly and to connect "inductively" to complex judgments. Already at the beginning of the seventeenth century, Bacon (1561–1626) called for a "modern science" that was radically delimited from medieval scholasticism and primarily oriented on perception and experience. The goal of the new science should be knowledge of nature that can be practically exploited. Berkeley (1685–1753) advocated an idealistic monism. He criticized Locke's distinction between primary qualities (extension, shape) and secondary qualities (color, smell, taste) in the perception of external objects. His consistent empiricism leads to the conclusion that only the spiritual and its contents and their experiences exist. He condenses his philosophy as follows: *Being is to perceive or being perceived. What is not perceived is not existent.*

2.3.2.1 Locke

John Locke (1632–1704) is considered the founder of empiricism. He argued critically with the medieval philosophy of being and stressed that only our senses establish the contact between us and the outer being. Locke distinguishes between the perception of external objects (sensation) and internal mental states (reflection). In this way, the contents of consciousness (ideas), such as thoughts, feelings, sensory images, memories, could be formed, which the human mind combines into ideas of things. However, we must analyze our mental capacity and find out what it is capable of and what it is not. This is the limit of what we can understand, what goes beyond this is not important, because it cannot penetrate us. In contrast to Plato's "innate ideas", Locke says that the human mind is a blank sheet (tabula rasa) at birth, which is described by experience in the course of life.

Fig. 2.9 The terms deduction and induction and their meaning

2.3.2.2 Hume

The British philosopher David Hume (1711–1776) asked how the contents of our consciousness come about. He states that all contents of consciousness are ultimately sensual perceptions in the form of impressions and ideas:

- Impressions are actual sensations when we see, hear, touch, etc.,
- Ideas are either reflections of impressions or associations of impressions due to similarities in spatial–temporal perception or due to cause-effect coherence.

Hume deduces that the "I" is merely a bundle of sensual perceptions and "things" are only series of perceptions in sensual consciousness. He doubts that "last certainty" is possible in the realm of experience.

- No causal necessity can be logically justified from experience.

Observations only allow the conclusion from experience that of similar causes similar effects can be expected with a certain probability. Hume took the position that an "induction" in the sense of a conclusion of individual observations to a general law is not permissible. No matter how many observations x with the properties y, it is not proven that the next observation x also has the property y. The negation of an "induction" implies consent to "deduction", under which, since Aristotle, the logical conclusion is understood from the general to the specific, Fig. 2.9.

2.3.3 Enlightenment

The Age of Enlightenment (also known as the Age of Reason) was an intellectual and philosophical movement that dominated the world of ideas in Europe during the 17th to 19th centuries.

2.3.3.1 Kant

Kant (1724–1804) overcame the opposing philosophical views of rationalism (primacy of understanding) and empiricism (primacy of sensual experience). His aphorisms on philosophy, ethics and the enlightenment are famous:

- Philosophy asks for the conditions of the possibility of experience.
- Ethics (*categorical imperative*): Act only according to that maxim by which you can at the same time want it to become a general law.
- Enlightenment is man's exit from his self-inflicted immaturity. Immaturity is the inability to understand his mind without the guidance of another. The enlightenment motto is: "Have the courage to use your own intellect".

Kant's philosophy deals with the structures of reason, i.e. the "apparatus of knowledge of human" for the recognition of things. Kant developed a *Transcendental Idealism* in his work *Criticism of Pure Reason* (1781) through the combination of rationalism and empiricism with the following basic ideas:

- All knowledge begins with sensual experience.
- A "thing in itself" means an "object" that exists, regardless of whether it is perceived by a "subject".

  According to Karl Jaspers, the "subject-object-split" is the irrevocable difference between the object of cognition (object) and the cognizing (subject).

- Cognition is structured by two types of determinations:

  - Variable (changeable, changing) determinations result from the constantly changing contents of cognition,
  - Invariable (unchangeable) provisions remain the same for all concrete findings.

- Transcendental is an epistemological reflection that is not related to the objects itself, but with the "conditions of the possibility of the kind of knowledge of objects" (not to be confused with the theological term transcendent).
- In order to gain knowledge, the two, not reducible from each other cognitive strains (a) sensuality, and (b) mind are required:

  - *Sensuality* is the ability to perceive the sensations of objects through a sensory apparatus. According to Kant, the form of contemplation (namely space and time) are present "a priori", all sensory impressions can only be the classification in space and time as source of knowledge.

    - *Space:* visual form of the "outer sense" (seeing, touching, etc.) for determination of the shape and size of objects and their relationship to one another.
    - *Time:* form of the "inner sense" (memory, imagination, judgment, inter alia) to determine the concurrence or succession.

  - *Mind* is the ability to form concepts and combine them to form judgments.

- Kant distinguishes four categories of basic judgment functions:

  - Quantity: unity, multiplicity, universality,
  - Quality: reality, negation, limitation,
  - Relation. e.g., cause-and-effect relation, interaction relation,

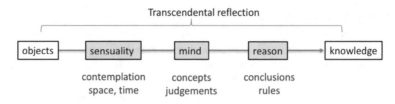

Fig. 2.10 The principle of transcendental reflection according to Kant in a simplified overview

- Modality: Possibility-impossibility, existence-non-existence, necessity-randomness.
- The term *reason* refers to the ability of human thought to draw conclusions, as well as to set up rules and principles.

A simplified scheme of Kant' philosophy with the trinity *sensuality – mind – reason* from the perception of an *object* to *knowledge* is shown in Fig. 2.10.

A commonplace example of transcendental reflection: Suppose I find an object in the street. With my senses of seeing and touching, I notice that there is a cylindrical, sharpened wooden stick with a black core. My mind says: this is a pencil and reason comes to the conclusion: I can write with it and in black letters.

With his epistemology, Kant answered his fundamental philosophical question "What can I know?" He was the first philosopher to emphasize that **human knowledge is not limited by what exists, but by our "knowledge apparatus"**. Everything we perceive in any way—be it an object, a feeling or a memory—we capture it with the help of our five senses, our brain and our central nervous system. What we cannot process with it, we cannot experience, because we can't capture it.

The epistemology of Kant sees as an indispensable condition for a valid acknowledge always the coming together of "view" and "thought" through the trinity *sensuality—mind—reason*.

Schopenhauer (1788–1860) followed with his eloquent work 'The World as Will and Representation' Kant thoughts and stated that Kant's philosophy reached "the decisive breakthrough in the history of the human mind". Schopenhauer developed Kant's ideas further and designed—with the involvement of reflections on the arts and eastern philosophies—a teaching that is equally includes ethics, metaphysics and aesthetics.

2.3.4   Cognition

In the occidental history of philosophy after to Kant, the occupation with the **Thinking** increasingly came to the fore. Two different views on the problem of thinking have emerged:

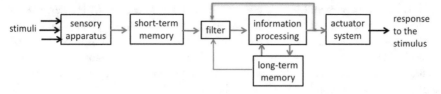

Fig. 2.11 The perceptual system as a model concept of cognitive science

- The theory of *behaviorism,* developed in the first half of the twentieth century in the context of psychology, excluded that mental operations, because of their "immateriality", could be the subject of scientific research. The thought process was considered to be an undetectable phenomenon (black box).
- In the 1960s, the *Cognitive Science* (science of the mind) made the thinking to its research field in an interdisciplinary approach, combining (a) the findings of cybernetics (science of control and regulation), (b) theories of the human psyche (psychology of thought), (c) the structure of the human brain and (d) neuroscience.

Cognitive science describes the human mind as a "system of information processing". An example is the thought process to trigger a human action as reaction to a stimulus received by a sensory organ (e.g., the reaction of a car driver to a traffic light signal).

The perceptual system as a model concept of cognitive science is illustrated in a simplified way in Fig. 2.11. Stimuli, emitted from the environment, hit the sensory apparatus as inputs and remain in the short-term memory for fractions of a second. From the multitude of the inputs, a filter makes a situation-specific selection through psychological operations. If the filter is closed, the stimuli are extinguished. If it is open, the stimuli enter the signal processing channel, where the actual information processing takes place in conjunction with long-term memory. The process takes place sequentially with a maximum of seven blocks of information, as has been demonstrated experimentally. The answer to the input stimulus is the response to the stimulus by an actuator system and the channel is emptied. Simultaneously with the emptying of the channel, a feedback, causes the re-opening of the filter thus enabling the start of a new processing cycle.

2.3.5 Existentialism

Existential philosophy is a direction of ego-philosophy, which sees the existence of man as the center. This development is historically seen as a turning point in philosophical thought from the object to the subject (the human being). Phenomenological experiments show that the correlation between the perceived object and the perceiving subject is influenced by intentions, prejudices, knowledge and beliefs. The existential philosophy was prepared by the phenomenological philosophy of Husserl (1859–1938). He coined the sentence: *I am—everything not-me is just "phenomenon" and solves into phenomenal contexts.* Kierkegaard (1813–1855) is considered the actual founder of existential philosophy. He says "Life can only be understood backwards—but it must be lived forwards". Objective thinking expresses everything in the result, but in subjective thinking everything is in motion. While in objective thinking, the thinking subject and his existence is indifferent, the subjective thinker as an existing person is interested in his own being. Existential philosophy sees two possibilities for this:

- Man can abandon himself and his personal-individual subjectivity and direct his interest to the objective facts only. That is the scientific point of view.
- But man can also, in subjective reflection, grasp his own individual being—the existence of the ego—as a possibility of radical freedom: I am what I conceive myself to be in existence. This is the position of existential philosophy.

The term *existentialism*, coined around 1930, encompasses philosophical views that in the "individual existence" see the fundamental characteristic of man. Existentialism is not so much a theory as a state of mind, which reflects the cultural climate after the Second World War. In theoretical terms, for existential philosophy, the difference between "objective and subjective reflection" is characteristic. Heidegger (1889–1976) poses the question of the "meaning of being", which necessitates to consider the existence of man as "one thing among other things", but to give it its own meaning of being. The philosophy of existence thus presents the human being with his own self as freedom and as a possibility.

But the responsibility for "free existence" also means leaving the everyday being-in-the-world and raises the central question about the "meaning of life". This question is discussed in existential philosophy with different approaches, here are only a few keywords:

- Kierkegaard saw the only solution in man's existential attachment to God.
- Jaspers (1883–1969) perceived the question of meaning of life and death as ciphers of God, which were assigned to people for interpretation.
- Bloch (1885–1977) expressed in his work 'The Principle of Hope' the expectation of a real democracy without "alienation" of people.
- Adorno (1903–1969) characterizes the question of the meaning of life as one of last questions in which the categories of metaphysics live on.
- Sartre (1905–1980) excluded God as the ultimate reason for meaning, whereas the absurdity of the existence follows, which man has to live through. Man is by chance "thrown into existence" after his birth and must actively try to give life meaning.
- In his book 'The Myth of Sisyphu's, Camus (1913–1960) compared being with the cycles of laboriously rolling a stone up a mountain.
- Sloterdijk (*1947) goes into the question of meaning in his book 'Critique of Cynical Reason' (1983) back to the cynicism/cynicism of antiquity. Cynicism is today, when people feel a great emptiness of meaning in their lives, however,

suffering from it suppresses. Their l ife will then be filled with practical constraints and the instinct of self-preservation.

2.4 Philosophy of Mind

The third fundamental area of philosophy, according to the Platonic Triangle, concerns the "philosophy of mind". This term comprises philosophical models in a synopsis of being (object) and I (subject), Ontology and Transcendental Philosophy.

2.4.1 The Absolute

The term *the Absolute* denotes being exempt from all restrictive conditions or relationships. On the basis of the meaning of this term—in particular in German idealism at the turn of the eighteenth and nineteenth centuries—various philosophical models developed to recognize "the whole of the world" and to establish a systematically structured school of thoughts. In the German idealism of the early nineteenth century, speculative views often dominated, which were not in line with the way of thinking of modern natural science, but had a rather aesthetic or mystical character. In some cases, new terminology and unusual (today difficult to interpret) word formations were introduced into philosophical models.

2.4.1.1 Fichte

Fichte (1762–1814) understood the absolute as "subjectivity and activity". The pure ego (the "I ") is a spiritual principle that underlies every reality. The Absolute creates the Nature, i.e. the material and passive reality. Fichte believed that people are capable of conscious action and are therefore "moral beings". It is the moral will and not the knowing spirit that is decisive for our human existence. He formulates in his science teaching three core tenets:

1. The ego sets its own being, establishes itself. The foundation of knowledge is not a fact, but a creative activity that is generated in the ego.
2. The ego opposes a non-ego (object). The opposed objects are in the conscious.
3. The I and the non-I are in a disjunction ("or-link").

According to Fichte, everything that can occur in the human mind must be derived from these principles.

2.4.1.2 Schelling

Schelling (1775–1854) saw in the absolute the indistinguishable unity of nature and spirit. Nature ("not the I") is a fundamental value, symmetrical to the mind and equally necessary. You can come to nature from the mind, but you can also go the other way around. Man is part of nature. Therefore, the human creativity is part of the productivity of nature. In man, nature has reached self-confidence. In his Berlin lectures Schelling presented a central question of the whole philosophy: Why is anything at all? Why is not nothing? This question is still regarded today as the ultimate question for anyone who does not believe in God.

2.4.2 Hegelism

Hegelianism is the philosophy of Georg Wilhelm Friedrich Hegel (1770–1831) which can be summed up by the dictum that "the rational alone is real". This means that all reality is capable of being expressed in rational categories. His goal was to reduce reality to a more synthetic unity within the system of absolute idealism. With his philosophy, Hegel aims to be able to understand the entire reality of the world in the diversity of its manifestations, including its historical development, in a coherent and systematic way. At the center of his system—in which he combined traditional metaphysics (Aristotle) and modern natural law (Locke)—stands "absolute", the "world spirit". The Absolute is for Hegel not immovable substantial, but dynamic, it develops "dialectically".

The Hegelian dialectic is not a formal technique of thinking. The "dialectical development" with its contrasts and contradictions is, in his view, a necessary spirit and concept and thus reality itself. The Hegelian dialectic consists of the "Three-step" of *thesis, antithesis, synthesis* and can be characterized in keywords as follows:

- The first moment is being itself (thesis).
- The second moment is the being outside of itself (antithesis).
- The third moment is the return to oneself (synthesis).

The infinity of knowledge, the "Absolute Mind" develops as a circular process, Fig. 2.12: As a thesis, the mind consists in the logic that expresses its rational being. As antithesis it is matter (philosophy of nature). As synthesis it consists of the process of a progressive spiritualization of matter (philosophy of mind).

In Hegel's system of philosophy there is nothing lasting, reality consists of a process of incessant becoming. No being can continue to exist by remaining the same as itself. (This was already emphasized by Heraclitus: *everything flows*). If the infinite is the finiteness of things, each fragment of reality has a specific value. In history, the stages of the progressive development of the mind (similar to gears in a mechanism) necessary and indispensable, there are in of history no positive or negative, legitimate or illegitimate events.

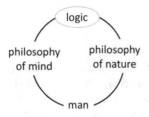

Hegel's encyclopedia of sciences
According to Hegel's philosophy, every aspect of reality is only explainable if it is seen in a dialectic circle in which the event participates, and which puts the event in relation to the rest of the world. Every thing has no reality per se, similar as an organ which cannot exist outside its body.

Fig. 2.12 The circular process of knowledge according to Hegel

According to Hegel's theory of justification, everything that happened has its precise and undeniable reason. If all finite things are included in a higher, global context, the logic of progression, the "Absolute" is reached. In Hegelian system of dialectical linking of *thesis, antithesis* as well as the abolition of opposites in a *synthesis*, a deductive linking along the lines of mathematical axiom systems was no longer considered.

2.4.3 Nietzsche

The poetic-linguistic work of Nietzsche (1844–1900) forms a contrast to the entire philosophy from Plato to Hegel. At its center is "life" as the subject par excellence, which is determined by the qualitative difference between two basic forces: one is the active, the other the reactive. The will to power in the difference of the forces makes all life a "fighting game".

Nietzsche claimed that God is death and considered the distinction between master and slave moralities. His works involved a sustained attack on Christianity and Christian morality, and he seemed to be working toward what he called the reevaluation of all values, and his position is often associated in the public mind with fatalism and nihilism.

Nietzsche emphasized a radical perspectivism, the view that perception, experience, and reason change according to the viewer's relative perspective and interpretation. It rejects the idea of one unchanging and essential world accessible to neutral representation by a disembodied subject. Perspectivism rejects objective metaphysics, claiming that no evaluation of objectivity can transcend cultural formations or subjective designations. Therefore, there are no objective facts, nor any knowledge of a thing-in-itself. Truth is thus created by integrating different vantage points together. Nietzsche's extraordinary reflections unfold their influence right up to contemporary philosophy.

2.4.4 Materialism

Materialism can be characterized as a basic direction of philosophy, opposite to idealism. The historical materialism of Marx (1818–1883) explains the functioning and development of human society from physical production. Marx adopted the method of dialectic from Hegel and kept Hegel's system as an abstract scheme, but replaced the "absolute spirit" by the "becoming of the material world". With this, in the doctrine of **Marxism** called after him, he puts Hegel's philosophical system "from head to toe". Instead of Hegel's "divine absolute", Marx declares the "material-economic absolute" of the production process or the work as the all-grounding reality. The fundamental model of historical materialism is the basic-superstructure diagram, Fig. 2.13:

- The ideological superstructure is the encroachment of the manifestations of the spirit (moral, ethical, religious, philosophical theories), underneath are the legal and political institutions, which refer to the mind (law, policy).
- The substantial basis is formed by the resources work and relations of production, with which each society ensures its material survival.
- the socialization of the means of production looked at.

For Marx, the material-economic basis is the supporting pillar of a society, since it forms the collective social production process of the human species. The predominant ideas in any society are those of the "ruling class", which hold in hand the levers of economic production through "capitalism". Marx understands his theory, which he developed together with Engels (1820–1895) in his Major work 'Capital' as "philosophy of the revolutionary proletariat".

As the "doctrine that eliminates domination", Marxism is the absolutely true doctrine to recognize and to expose teachings, philosophies, and religions as outdated ideologies. With the further development of the productive forces, the proletariat should emancipate and in the last revolution eliminate capitalism in order to create a "classless Society". As a prerequisite for a classless society, Marxism becomes the end of the exploitation of man by man, the abolition of private property.

As version of an "undogmatic Marxism", the *Critical Theory of Society* (Frankfurt School) is denominated. It was conceived by Horkheimer (1895–1973) and Adorno

Fig. 2.13 Concept scheme between the Marxian terms Base and Superstructure

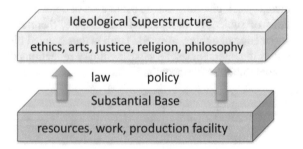

(1903–1969) as an analysis of capitalist relations of production and at the same time as a theory of rationality in general. According to Horkheimer, Critical Theory has "people as the producers of all their historical life forms" as its content. For Habermas (* 1929) form "communicative interactions with rational reasons for validity" is the basis of society.

After the dogmatization of the Marxist theses by Lenin (1870–1924), Marxism-Leninism at the beginning of the twentieth century had a strong ideological effect. After political upheavals (e.g. the October Revolution of 1918 in Russia), the doctrines of the "States of Real Socialism" evolved. However, the (Party) leaders of the propagated "classless society" (e.g. Stalin, Mao) became despots, and the so-called "socialist parties" became the "Ruling Classes" in the socialist states. At the end of the twentieth century, under protest movements of the populations, the ideological Marxism, which has led to "state absolutism", broke down.

2.5   Analytical Philosophy

The term *analytical philosophy* summarizes various philosophical trends that were developed at the end of the 19th and in the twentieth century. Comte (1798–1857) justified a direction of philosophy entitled "positivism" by rejecting metaphysics: "Instead of looking for ultimate causes, philosophy must be deeply rooted in facts and rules". According to his three-stage doctrine, the development of mankind proceeds in three stages: (1) the theological, (2) the metaphysical and (3) the positivist stadium. The positivist stage is the result of progressive thinking. With that, mankind has overcome religious and metaphysical superstition and reached the level of scientific knowledge.

The scientific philosophers of positivism consider most of philosophy from Plato to Hegel as "unscientific". For them, the exact natural sciences are the sciences par excellence. All others sciences should be, in the sense of the method of the exact natural sciences, transformed to a single unified science. The basic concern is to constitute the system of sciences from only two elements:

- empirical (sensory) elementary experiences and
- their formal-logical links.

The different perspective of the „things of the world " by the different philosophical views is made clear by the following keywords:

- The classical philosophy of metaphysics asks for the conditions of cognition of the things of the world: Why are there things at all?
- The Transcendental Philosophy of Kant deals with the structures of reason, i.e. the "cognitive apparatus of man" for recognizing things.
- Analytical philosophy does not examine "things in themselves", but analyses the logic and the language of how things are spoken.

Fig. 2.14 An aphorism for
the impact of language

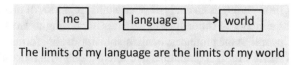

The limits of my language are the limits of my world

2.5.1   Logical and Linguistic Analysis

The formal logic, which is fundamentally important for analytical philosophy, was developed by Frege (1848–1925). In it symbols and rules as well as in its methods for the combination of symbols and rules, valid conclusions are specified.

Peano (1852–1932) shows that mathematical statements cannot be accepted by intuition, but must be derived from axiomatic premises.

With the *Principia Mathematica*, a three-volume work published in 1913 on the foundations of mathematics, Whitehead (1861–1947) and Russel (1872- 1970) attempted to derive all mathematical truths from a well-defined set of axioms and rules of symbolic logic. Russel developed an "abstract cosmology", which deals with the last structures of language and the world. The goal is to make sentences logically transparent in such a way that they only contain that with which one is directly acquainted (e.g. sensory impressions or logical links). What is not "known" cannot be "named". Based on Russel's linguistic analysis,

Wittgenstein (1889–1951) developed his Theory of Representation. According to his model, the world consists of things and their "configurations", the "facts of the case". Things form the "substance" of the world; they are as such simple, unchangeable and independent of facts of the case. In a fact of case, things are linked by a relation. The general form of a fact of case is "aRb", i.e. "a is in a relationship with b." The relations form the logical framework of the world and thus also the common ground of language and the world. The certain language, in which we are experiencing the world, also determines our view of the world, Fig. 2.14. The boundaries of the describable are the "boundaries of the world".

2.5.2   Epistemology

The philosophy of the twentieth century has intensively dealt with the question of verification - the proof of the accuracy of presumed or alleged facts. Thereby, "induction", the method of inferring a general conclusion from observed individual phenomena is not considered permissible. This led to the fact that the classical metaphysics, which claims to recognize the things of the world independently of experience, is rejected. Critical rationalism holds the view that philosophy first of all has to be the theory of scientific knowledge.

Popper (1902–1994) develops the method of *falsification*, under which the proof of the invalidity of a statement, method or hypothesis is understood. Accordingly, a hypothesis cannot be "verified" but only "falsified." Popper explains the theory of falsification with a simple example. Suppose, the hypothesis is: "All swans are white". The finding of numerous white swans only contributes to the hypothesis and may be maintained. There is always the possibility to find a swan of a different color. If this case occurs, the hypothesis is disproved. As long as no swan of a different color was found, the hypothesis can still be considered as being not refuted.

According to Popper, the method of falsification is an important instrument of epistemology. Physical reality exists independently of the human mind. Therefore, it can never be directly grasped. To explain it, we provide plausible theories, and when these theories are practically successful, we apply them. If a theory proves inadequate, we look for a better, more comprehensive theory without the identified limitations. Ultimately, theories cannot generally be "proven", but they can already be rebutted by a single counter-observation (e.g. a black swan in the above example). As in the light of current knowledge, the truth of a philosophical model cannot be proven, the "striving for certainty", in which numerous philosophical models were biased, has to be abandoned. A scientific theory must allow the possibility of being refuted (falsified) by an observation. If this is not allowed by the representatives of the theory, it is not a scientific theory but an "ideology".

2.5.3 Limits of Insight

In the twentieth century, numerous positions of ancient philosophy were rethought, whereby knowledge was expanded but also limits of knowledge became visible. This is illustrated by the following examples from the fields of logic, geometry and in general of mathematics.

Logic: The classical logic of Aristotle postulates that a statement can only be either true or false. In further development of Aristotelian logic, Boole (1815–1864) created a "binary" propositional logic by using only the numbers 1 and 0, which were later applied in computer languages. An extension of the logic founded by Aristotle, was created in 1965 by Zadeh with the "Fuzzy Logic". The fuzziness marking of objects is specified gradually using numerical values between 0 and 1. This allows the "fuzziness" of specifications such as "a little", "quite", "strong" or "very" in mathematical models which are evaluated quasi-logically.

Geometry: The descriptive geometry of the plane or three-dimensional space is supported by the postulates and axioms of Euclid (360–280 B.C.), which are considered valid for more than 2000 years. Of particular importance for the geometry of the plane is Euclid's fifth postulate (parallel axiom): *In a plane, given a line and a point not on it, at most one line, parallel to the given line can be drawn through the point.* Lobachevsky (1793–1856) and Riemann (1826–1866) developed mathematical models of geometric spaces in which the parallel axiom has no validity. Riemann started from a contrary postulate: a "plane" is conceivable in which there are no parallels, i.e. two straight lines in the plane always have one point in common.

With this postulate he developed the "elliptical geometry" which describes a world, in which the geometric surface curves and takes on the shape of a sphere. In a generalized form, Riemannian curvature is the curvature of any dimensions. It has a central role in four-dimensional space–time of Einstein's theory of relativity.

Mathematics: Mathematics has been the ideal model for all other realities since antiquity and Plato's theory of ideas. In the 1920s, the mathematician David Hilbert (1862–1943) had proposed the *Hilbert Program,* named after him, which aimed to prove the "freedom of mathematics from contradictions". The aim was to find a strictly formalized calculus with simple, immediately plausible axioms in order to place mathematics and logic on a common, consistent basis. This should make it possible to prove for every mathematical theorem whether it is true or false. All true propositions should be derivable from the new axiom system. For this it would have to be "free of contradictions" and "complete". While for some parts of mathematics, such as *number theory,* the absence of contradictions could be determined (Gentzen, 1936), this could not be shown for the entire mathematics.

Gödel (1906–1978) proved in 1931 that there are contradictory statements in mathematics—especially in the set theory of Cantor (1845–1916) and in the Principia Mathematika of the philosophers Whitehead and Russel. The Gödel incompleteness theorems refer to formal systems of chains of symbols and rules and are.

- Any sufficiently powerful formal system is contradictory or incomplete.
- Any sufficiently powerful formal system cannot prove its own consistency.

This means that a mathematical system (such as arithmetic) has basic statements that it cannot prove itself. Mathematics cannot be grasped by a "structure without contradiction". Thus, **a reference to mathematics and the application of mathematical analogies cannot "prove" the "truth" of philosophical statements**.

2.6 Three-World Theory

At the beginning of the twentieth century, the philosopher and mathematician Frege wrote a three-part theory, in which he stated: besides the realm of objective-real physical objects (1), and (2) the realm of subjective *conceptions*, there exists (3) the realm of objectively-non-real thoughts. All they are grasped by the consciousness but not brought forth.

Three-part (trinity) models have been developed several times in the history of philosophy, for example.

- In Greek antiquity as *Logos—Psyche—Physis*,
- With the Romans as *Ratio—Intellectus—Materia*,
- In the philosophy of Kant as *Sensuality—Mind—Reason*.

Popper's Three-Worlds-Doctrine is a model that supposes the existence of three "worlds", Fig. 2.15:

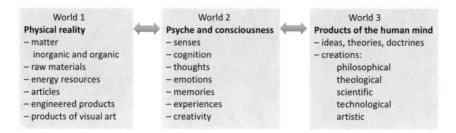

Fig. 2.15 The model of Popper's three-world theor

- World 1 is the totality of material objects. It consists of physical facts and matter of all kinds—including the human body and brain—and of all objects, works of art and material products.
- World 2 comprises the mental disposition and the totality of states of consciousness and thought processes of humans. This concerns the sensations of the visual, acoustic and haptic sense, soul states, thoughts, memories, emotions and human creativity.
- World 3 represents the knowledge and products of the human spirit, which theories, ideas and result contents. They represent the impersonal world of intellectual services of mankind and are available like "objects".

In the Three-Worlds-Doctrine, Popper argues that all three worlds are real, because causal interactions could be observed. Together with the brain researcher and Nobel Prize winner John Carew Eccles (1903–1997) Popper developed a simple model, where the "brain was compared with a computer" and the "I was compared with its programmer".

According to new findings in brain research, thinking is based on combined electronic and chemical activities of nerve cells (neurons). Our brain has billions of neurons and the number of potential neuron combinations is greater than the number of atoms in the universe. The neurons respond to certain characteristics and synchronize in different bilateral connections between individual cells, which are either genetically designed or formed by a learning process. When a certain pattern of characteristics appears, a Neural network of interconnected nerve cells is formed. For each feature pattern, a different neuron ensemble is aroused together. The neural networks form the structure and information architecture of the brain. They are thus the basis for psychic disposition and states of consciousness (world 2) which, according to Popper, acts as a mediator between world 3 of the products of the human mind and world 1 of the material reality of physical objects.

Chapter 3
Physics

Physics explores and describes nature and the laws of nature. The chapter presents the essentials of physics and describes the world view of physics.

- *Physics tries to find out details of natural phenomena through experiments, to observe objectively and to understand their legality. It strives to formulate the connections mathematically and thus to create laws which are valid in the whole cosmos without restriction. Finally, this made it possible to utilize the forces of nature in technology to serve our purposes* (Werner Heisenberg).
- *Only the knowledge of the laws of nature allows us to turn the sensual impression into the underlying process* (Albert Einstein).
- *Observation, reason and experiment make up what we call the scientific method. The test of all knowledge is experiment. Experiment is the sole judge of scientific truth* (Richard P. Feynman).

It should be noted, however, that physics cannot directly recognize the "nature in itself", i.e. the "untouched" nature. *The physical world cannot be grasped directly with the human mind* (Popper). To be able to "see" objects, we must "touch" them with mechanical, electromagnetic or optical probes in order to perceive them physically. When we observe macroscopic objects of our daily experience, the observed object is practically not changed by the physical process of observation.

In the case of the sub-microscopic building blocks of matter and elementary particles, however, the observation process can represent an "intrusion" into the object or a "disturbance", so that one can no longer speak here of normal particle behavior detached from the observation process. Heisenberg sees this as follows in his book "The picture of nature in today's physics" (1955):

In the natural sciences, the object of research is no longer only nature itself, but nature exposed to the human question. The laws of nature, which we formulate mathematically to describe the elementary particles of matter, no longer deal with the elementary particles themselves, but with our knowledge of them. If we can speak of an image of nature in the exact natural science in our time, then it is actually no longer an image of nature, but rather an image of our relations to nature.

© The Author(s), under exclusive license to Springer Nature Switzerland AG 2021
H. Czichos, *The World is Triangular*,
https://doi.org/10.1007/978-3-030-64210-5_3

For the description of physical phenomena some elementary terms have been coined. The terms of physics for physical objects are verbal characterizations and are not defined as "sharply" as the terms of mathematics for abstract objects. The following terms have a central meaning in physics:

- **Space** has, according to experience, three dimensions (height, width, depth), it can contain matter and fields, and all physical processes take place within it.
- **Time** characterizes the irreversible sequence of physical processes. In the theory of relativity, space and time merge into a uniform space–time, whose structure depends on the distribution of matter and energy.
- **Forces** are directed physical quantities (vectors). They can accelerate or deform bodies, as well as perform work or change the energy of a body through force effects.
- **Particles** are objects that are small compared to the scale of the system under consideration; they can be regarded as point-like centers of force fields.
- **Fields** describe the spatial distribution of a physical quantity in space. They can themselves be regarded as physical objects and possess momentum and energy (field energy).
- **Oscillations** are time-periodic processes.
- **Waves** are time-periodic processes which—as mechanical waves in media or as electromagnetic waves in a vacuum—propagate spatially and transfer energy or transport information. Wave- packets can be attributed particle properties(wave-particle duality). Waves of matter characterize the probability of moving particles in space.
- **Bodies** are objects that have mass and occupy space.

3.1 Physics of Matter

The fundamental importance of matter for the world of physics was illustrated by Feynman in his famous LECTURES ON PHYSICS with the question: *If, in some cataclysm all of scientific knowledge were to be destroyed, and only one sentence passed on to the next generations of creatures, what statement would contain the most information in the fewest words?* According to the Nobel Prize winner in Physics, this is the sentence: *"All things are made of atoms"*.

The atomic structure of matter—which was postulated by Democritus around 400 BC, (see Sect. 1.2.1) could be made "visible" experimentally in 1951 by Erwin W. Müller using the field ion microscope, Fig. 3.1. The experimental apparatus consists of a metal tip (object) in a noble-gas filled, deep-frozen vacuum chamber (to reduce thermal movement of the atoms of the metal tip to be imaged) and a high voltage between the object and a fluorescence screen. The atomic structure is visualized as follows: When a noble gas atom (helium or neon) hits an atom of the metal tip, it is positively charged electrically and accelerated in the electric field of the high voltage towards the screen, where it creates an image point. The pattern that the dots form on the screen is the central projection of the atomic structure of

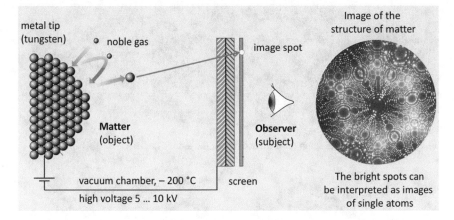

Fig. 3.1 Principle of the field ion microscope to visulize the atomic structure of matter

the metal tip. This made it possible for the first time, that people can "see" atoms magnified millions of times.

The extreme smallness of atoms is illustrated by the following comparison, Fig. 3.2: *If you enlarge an apple to the size of the Earth, the atoms of the apple are about the natural size of the apple* (Feynman).

The discovery that atoms have an atomic shell and an atomic nucleus was the result of experimental examination of atoms with radioactive radiation (Ernest Rutherford, 1911). According to the simplified illustration in Fig. 3.3a, the experimental setup consists of a radiation source that emits alpha particles that form an image point on a screen. If a thin gold foil (Au atoms) is introduced into the beam, a deviation of the beam can be observed, Fig. 3.3b. From the observed deflection angle it can be concluded that the deflection is due to the positive charge cluster which is concentrated in a very small "atomic nucleus" which is only a fraction of the atomic shell.

The experimentally determined size ratio of the dimensions of the atomic nucleus to the atomic shell is about 1:10,000 and corresponds in a striking comparison to the size ratio of the diameter of a golf ball (atomic nucleus) to the height of the Eiffel Tower (atomic shell), Fig. 3.4.

Fig. 3.2 The size of an atom is related to the size of an apple in the same way as apple size is related to the size of the earth

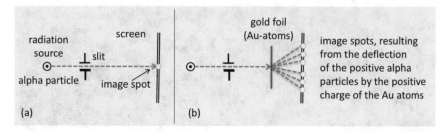

Fig. 3.3 The experimental setup for the discovery of the atomic nucleus

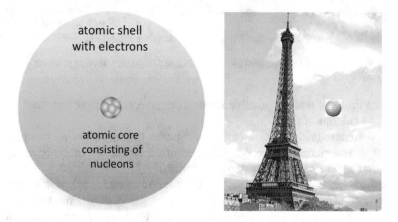

Fig. 3.4 A comparison between atomic and macroscopic proportions

3.1.1 The Structure of the Atom

In simplified representation, an atom consists of an atomic nucleus (ø approx. 10 femtometers) with nucleons (positively charged protons p and neutral neutrons n), which make up 99.9% of the atomic mass and an atomic or electron shell (ø approx. 0.1 nm) with negatively charged electrons (carriers of the elementary electric charge e). Protons and neutrons are composed of up quarks u and down quarks d (proton: $p = 2u + d$, neutron: $n = 2d + u$). Atoms can be modelled by Bohr's atomic model ("planetary model" with electrons on discrete orbits around the atomic nucleus), Fig. 3.5, and quantum mechanically by the orbital model (electrons as three-dimensional waves with probability distributions for location and momentum).

The **shell model of atomic physics** is an extension of Bohr's atomic model and a simplification of the orbital model: electrons "circle" around the atomic nucleus in "shells", designated K (max. 2 electrons), then L (max. 8 electrons), and so on. The location of the electrons is described by a probability function (Schrödinger equation). The state of each electron is determined by four quantum numbers: Primary quantum number (denotes the shell K, L, M, …), secondary quantum number (denotes

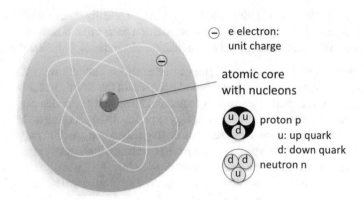

Fig. 3.5 The atomic model of matter

side shells), magnetic quantum number, spin quantum number. Since, according to the Pauli principle, no two electrons are identical in all four quantum numbers, the electrons are distributed among the different quantum-mechanically permitted states of the main and secondary shells. In the neutral state, electron charge and nuclear charge are equal, electrically charged atoms are called ions. The structure of the electron shell largely determines the physical properties of the atoms. For example, the "origin of light" can be explained by the emission of light quanta (photons) resulting from quantum jumps of electrons between different orbitals, Fig. 3.6.

The **shell model of nuclear physics** is a model of the structure of atomic nuclei. It considers the individual nucleons and their movement according to the rules of quantum mechanics, similar to the electrons in the atomic shell. Elementary particle-physics and high-energy physics have developed from nuclear physics. Radioactivity is the property of unstable (or artificially made unstable) atomic nuclei to transform, while releasing energy:

- *Alpha radiation* is ionizing particle radiation (helium atomic nuclei) with a low penetration depth into matter.
- *Beta radiation* is ionizing radiation (electrons) with biological effects, which is dependent on the radiation energy and duration of radiation.

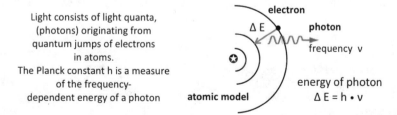

Fig. 3.6 The principle of the generation of light

- *Gamma radiation* is material-penetrating and biologically damaging electromagnetic radiation, lead shielding serves as radiation protection. It is used in material testing (radiographic testing) and in medicine (diagnostics, radiation therapy).

The analysis of the atomic structure of matter shows that atoms consist of three long-term stable components: protons and neutrons in the atomic nucleus and electrons in the atomic shell. Sub-nuclear (unstable) elementary particles can be generated with particle accelerators in high-energy experiments. According to the *standard model of elementary particle physics* it is distinguished in a special "particle terminology" six quarks, six leptons, several bosons (exchange particles) and the so-called Higgs boson, which was experimentally detected in July 2012 with the Large Hadron Collider at the European research center CERN. The physicists Higgs, Englert and Akasaki were awarded the Nobel Prize for Physics in 2013 for their theoretical prediction of the Higgs particle, which contributes to the understanding of the origin of the mass of subatomic particles. A further labeling of elementary particles is not given here.

- *If one wants to talk about elementary particles, one must either use a mathematical scheme as a supplement to ordinary language or one must combine it with a language that uses a modified logic or no well-defined logic at all (Werner Heisenberg).*

3.1.2 Aggregate States of Matter

Types of atoms with the same nuclear charge number (atomic number) are called chemical elements. A neutral chemical element has the same number of electrons and protons. Atoms with the same number of protons but different numbers of neutrons are called isotopes. Atoms with completely occupied shells are noble gases. With the other elements, the outer region of the electron shell (valence shell) is responsible for the chemical bonds and the formation of molecules or crystalline structures. The 92 natural chemical elements, ordered in the Periodic Table (lightest element: hydrogen H, heaviest element uranium, U) form the basis of chemistry. Matter has three states of aggregation: gaseous, liquid, solid, an example is shown in Fig. 3.7.

- **Gases** are substances whose particles move freely; they completely fill the space available to them. The microscopic kinetic energy E of the gas particles determines

Water molecule with the atoms hydrogen (H) and oxygen (O), observable in nature in three aggregate states

water vapor   water   ice

Fig. 3.7 Illustration of the three states of matter using water as an example

the macroscopic gas temperature T according to $E = kB - T$, where kB is the Boltzmann constant (natural constant). The number of gas particles per volume unit of an ideal gas is a physical constant (Loschmid constant N_L. Reference value for the amount of substance is the mol. 1 mol of a substance contains as many particles as there are carbon atoms in 12 g of carbon ^{12}C. The number of particles per mole is called Avogadro constant N_A, it has the dimension of a reciprocal amount of substance and is a natural constant.

- **Liquids** are volume-resistant but shape-unstable substances that exert hydrostatic pressure on container walls. They are described microscopically due to their constant temperature-dependent movement (Brownian motion) by means of statistical mechanics and macroscopically by continuum mechanics and fluid mechanics.

- **Solids** have a regular arrangement of their atomic or molecular components in crystallized form (long-range order). In an amorphous arrangement, bonds exist only between the closest neighbors (near order). From the view of Solid States Physics the fundamental features of a solid material are:

- Atomic nature of solids: The atomic elements of the periodic table, which constitute the chemical composition of a material.

- Atomic bonding: The type of cohesive electronic interactions between the atoms (or molecules) in a material, empirically categorized into the following basic classes:

  - Ionic bonds form between chemical elements with very different electron negativity (tendency to gain electrons), resulting in electron transfer and the formation of anions and cations. Bonding occurs through electrostatic forces between the ions.

  - Covalent bonds form between elements that have similar electron negativities, the electrons are localized and shared equally between the atoms, leading to spatially directed angular bonds.

  - Metallic bonds occur between elements with low electron negativities, so that the electrons are only loosely attracted to the ionic nuclei. A metal is thought of as a set of positively charged ions embedded in a sea of electrons.

  - Van der Waals bonds are due to the different internal electronic polarities between adjacent atoms or molecules, leading to weak (secondary) electrostatic dipole bonding forces.

- Spatial atomic structure: The amorphous or crystalline arrangement of atoms (or molecules) resulting from "long range" or "short range" bonding forces. In crystalline structures, it is characterized by unit cells which are the fundamental building blocks or modules repeated many times in space within a crystal.

- Grains: Crystallites made up of identical unit cells repeated in space, separated by grain boundaries.

- Phases: Homogeneous aggregations of matter with respect to chemical composition and uniform crystal structure: grains composed of the same unit cells are the same phase.

- Lattice defects: Deviations of an ideal crystal structure:

- – Point defects or missing atoms: vacancies, interstitial or substituted atoms
- – Line defects or rows of missing atoms: dislocations
- – Area defects: grain boundaries, phase boundaries, twins
- – Volume defects: cavities, precipitates

- • Microstructure: The microscopic collection of grains, phases, and lattice defects.

3.1.3 Microstructure of Materials

Materials result from the processing and synthesis of matter, based on chemistry, solid state and surface physics, influenced also by the engineering component design and production technologies. The microstructure of materials is illustrated for a metallic material in Fig. 3.8 and for a polymeric material in Fig. 3.9.

3.2 Physics of Space, Time and Gravity

All physical phenomena occur in space and time. According to our present knowledge. the space of physical phenomena extends from the sub-nanoscopic elementary particles up to the border galaxies of the universe at a distance 45 billion light years

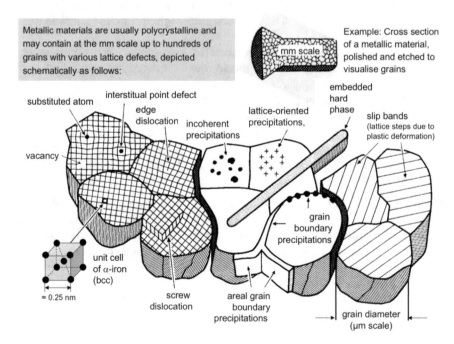

Fig. 3.8 Illustration of the microstructure of metallic materials

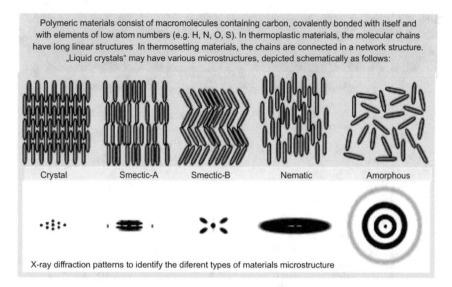

Polymeric materials consist of macromolecules containing carbon, covalently bonded with itself and with elements of low atom numbers (e.g. H, N, O, S). In thermoplastic materials, the molecular chains have long linear structures In thermosetting materials, the chains are connected in a network structure. „Liquid crystals" may have various microstructures, depicted schematically as follows:

| Crystal | Smectic-A | Smectic-B | Nematic | Amorphous |

X-ray diffraction patterns to identify the diferent types of materials microstructure

Fig. 3.9 Illustration of the microstructure of polymeric materials

from Earth. One light year denotes as astronomical measure of length the distance that light travels in space in one year, that is 9.46 trillion kilometers (9.46×10^{12} km). The time scale ranges from the transit time of light through an atomic nucleus (10^{-24} s) to the age of the earth of about five billion years and the age of the universe of about 14 billion years. An overview of the dimensions of space and time is given in Table 3.1. The question marks designate the limits of today's experimentally verifiable knowledge.

3.2.1 Space and Time in Antiquity

In antiquity, space is understood as a kind of "container" for all things of the cosmos. It is assumed that every body occupies a place and also every place is always occupied by a body. (Where one body is, there can be no other).

Time was historically understood as a concept independent of space. It describes the sequence of events that people perceive as past, present and future. The concepts of space and time in antiquity were developed by Aristotle; they are presented in Sect. 2.2.5.

Table. 3.1 Overview of the dimensions of space and time

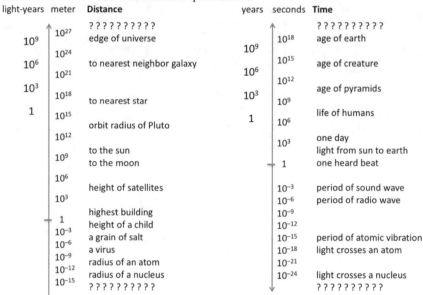

light-years	meter	Distance	years	seconds	Time
		? ? ? ? ? ? ? ? ? ?			? ? ? ? ? ? ? ? ? ?
10^9	10^{27}	edge of universe		10^{18}	age of earth
	10^{24}		10^9		
10^6		to nearest neighbor galaxy		10^{15}	age of creature
	10^{21}		10^6		
10^3	10^{18}			10^{12}	age of pyramids
		to nearest star	10^3	10^9	
1	10^{15}				life of humans
		orbit radius of Pluto	1	10^6	
	10^{12}				one day
		to the sun		10^3	light from sun to earth
	10^9	to the moon		1	one heard beat
	10^6				
		height of satellites		10^{-3}	period of sound wave
	10^3			10^{-6}	period of radio wave
		highest building		10^{-9}	
	1	height of a child		10^{-12}	
	10^{-3}	a grain of salt		10^{-15}	period of atomic vibration
	10^{-6}	a virus		10^{-18}	light crosses an atom
	10^{-9}	radius of an atom		10^{-21}	
	10^{-12}	radius of a nucleus		10^{-24}	light crosses a nucleus
	10^{-15}	? ? ? ? ? ? ? ? ? ?			? ? ? ? ? ? ? ? ? ?

3.2.2   Space and Time in Classical Mechanics

The classical concepts of space and time were presented by Isaac Newton in his work
Philosophia Naturalis, Principia Mathematica (1687) as follows:

- The absolute space remains by its nature and without relation to an object always
  constant and immovable.
- The absolute, true and mathematical time proceeds by itself and by its nature
  uniformly and unrelated to any external object.

Based on these definitions, Newton formulated the basic laws of classical
mechanics. A coordinate system in which Newton's axioms apply is called an *inertial system*. In inertial systems, the laws of classical mechanics have the same form
(Galilean principle of relativity).

- First law: every body persists in its state of rest or uniform motion unless it is
  forced to change its state by forces acting on it.
- Second law: the change of motion is proportional to the action of the moving force
  and occurs according to the direction of the straight line in which that force acts.
- Third law: forces always occur in pairs. If a body A exerts a force on another body
  B (actio), a force of the same magnitude but opposite direction is exerted by body
  B on body A (reactio).

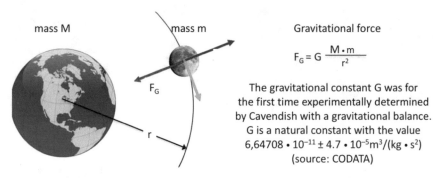

mass M   mass m   Gravitational force

$$F_G = G \, \frac{M \cdot m}{r^2}$$

F_G

r

The gravitational constant G was for
the first time experimentally determined
by Cavendish with a gravitational balance.
G is a natural constant with the value
$6{,}64708 \cdot 10^{-11} \pm 4.7 \cdot 10^{-5} \, \text{m}^3/(\text{kg} \cdot \text{s}^2)$
(source: CODATA)

Fig. 3.10 Illustration of Newton's gravitational force

It follows from Newton's axioms of classical mechanics that every mass-bearing body acts instantaneously throughout space on every other body with an attracting gravitational force. The gravitational force acts along the linear connecting line of both bodies; its magnitude is proportional to the product of the two masses and inversely proportional to the square of their distance, see Fig. 3.10.

The axioms of classical mechanics give a common explanation for space and time and gravity on the earth, as well as for the moon's orbit and the planetary movements around the sun. The entire technical mechanics of mechanical engineering, civil engineering and aeronautical engineering is also based on Newton's principles of classical mechanics.

3.2.3 Fields in Space and Time

Newton's theory of gravity is a so-called "theory of remote action". It does not explain how the gravitational interaction between two distant bodies A and B is transported through space, whereby the propagation speed of the interaction would have to be unlimited. After the effects of electricity and magnetism and the electromagnetic force between electrically charged particles were discovered, Faraday developed the concept of the electromagnetic "field."

As already mentioned in the introduction, in physics, fields describe the spatial distribution of a physical quantity in space. They can be considered themselves as physical objects:

- Fields fulfill equations of motion (field equations). The dynamics of fields are dealt with in field theory. For the electric and magnetic field, the Maxwell equations are the equations of motion, they are described in Sect. 3.7.5.
- Like bodies, fields have energy (field energy), momentum and angular momentum. The force action between two bodies in space is explained by the fact that a field absorbs these quantities from one body and transfers them to the other body.

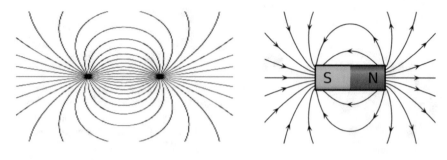

Electric field lines illustrate the
transmission of force between two
electrically charged particles

Magnetic field lines illustrate the
transmission of force between the
poles of a magnet

Fig. 3.11 Illustration of the electric and magnetic field

Figure 3.11 shows experimentally representable examples of electric and magnetic fields in space.

Electromagnetic fields form the physical basis for all telecommunications: Radio, television, telephone, computers, satellite navigation systems, WLAN, Internet.

3.2.4 Relativity of Space and Time

The Special Theory of Relativity (SRT) describes the behavior of space and time from the point of view of observers moving relative to each other, it is presented in Sect. 3.7.2. In classical mechanics, it is assumed that all events take place in the three-dimensional space of Euclidean geometry and that the laws of mechanics apply equally in every inertial system, i.e. in every system that is not accelerated (principle of relativity, Galilean transformation of reference systems). At the end of the nineteenth century it was recognized that the Maxwell equations, which describe electrical, magnetic and optical phenomena are not Galilean-invariant. Furthermore, it was experimentally established that the speed of light, contained in the Maxwell equations, has a constant value of c ≈ 300,000 km/s for both stationary (resting) and dynamic (moving) senders, Fig. 3.12.

Speed of light is the velocity at which all
massless particles and field perturbations
travel in vacuum, including electromagnetic
radiation and gravitational waves

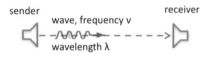

speed of light $c = \lambda \cdot v = 299.792.458$ m/s

Fig. 3.12 Observation of the speed of ligt

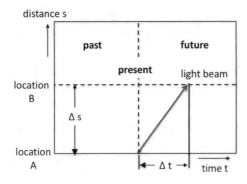

Light has a velocity of $c \approx 300.000$ km/s. As $c = \Delta s / \Delta t$ and $\Delta s = c \cdot \Delta t$, a light beam with an information from location A (present) reaches a location B (future) only after the time Δt.
Example: If between A and B is distance of 300 m, an event at A is recognized at B with a delay of 1 millisecond, i.e. the event at A is not "simultaneously" recognized at B.

Fig. 3.13 The relativity of space and time and the concept of "simultaneity"

In order that the Maxwell equations also apply unchanged in every reference system, the Lorentz transformation (see Sect. 3.7.2) had to be used instead of the Galilean transformation. The Lorentz transformation is not a pure transformation of the three-dimensional Euclidean space like the Galilean transformation, but a transformation in a mathematical four-dimensional space–time continuum (Minkowski space). From these considerations it follows that it makes sense to regard space and time as one unit in Minkowski space—similar as the unit of length–width-height in three-dimensional Euclidean space. The four-dimensional unit from space–time is called space–time. Accordingly, the basic equations of classical mechanics had to be reformulated because they are not Lorentz-invariant. For low speeds (car, plane, rocket) however, the Galilean and Lorentz transformations are so similar that the differences cannot be measured, i.e. the relativity of space and time is not perceptible in everyday life.

In the theory of relativity, the term "simultaneity" is put into perspective. An event at a location A does not take place "simultaneously" for an observer at location B, but only after the time it takes light to travel from A to B. For example, a "sunrise" currently observed on earth (location B) is already past for the sun (location A), because light needs about 8 min for the distance sun–earth. In general, "present" at a location A is "future" at a location B, and "present" at location B is "past" at location A, see Fig. 3.13.

3.2.5 Space, Time and Gravity

The connection between space, time and gravity is described by Einstein in the General Theory of Relativity (ART). He formulates ART by comparing Newtonian gravity with Faraday's concept of the electric field. He assumes that, just like electricity (Fig. 3.8), gravity (Fig. 3.7) must also be supported by a field. From this comparison, Einstein developed the basic idea of general relativity, which the theoretical physicist Carlo Rovelli describes as follows in his book *Seven Short Lessons in Physics* (Rowohlt 2016):

- *The gravitational field is not spread out in space, but it is space. Newton's space, in which things move, and the gravitational field as a carrier of gravity are one and the same thing.*

The mathematical basis for the description of space–time in ART is Riemann's geometry. Riemann developed elliptical geometry, which describes a world in which the two-dimensional geometric surface curves and takes on the shape of a sphere. In generalized form, Riemann's curvature is the curvature of any dimensions. For four-dimensional space–time, the same mathematical tools are used as for the description of a two-dimensional spherical surface. In the ART, Riemannian geometry replaces the Euclidean geometry of the "flat" three-dimensional space of classical mechanics. The "curvature" of space and time is caused by any form of energy, such as mass, radiation or pressure. These quantities enter into the Einstein equations of ART as the source of the gravitational field. The resulting curvilinear motion of bodies is attributed to gravitational acceleration. Figure 3.14 shows model representations.

The Einstein field equations of the ART indicate how the matter and energy content affects the curvature of space–time. They also contain information about the effect of space–time curvature on the dynamics of particles and fields. From the ART equations, Einstein developed a series of predictions (which at first seemed improbable), all of which were experimentally confirmed:

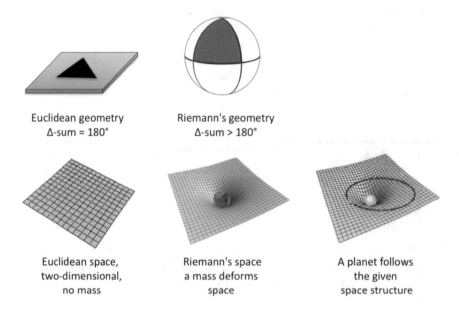

Euclidean geometry
Δ-sum = 180°

Riemann's geometry
Δ-sum > 180°

Euclidean space,
two-dimensional,
no mass

Riemann's space
a mass deforms
space

A planet follows
the given
space structure

Fig. 3.14 Model representations of gravity and planetary motion in a gravitational field

- General relativity provides the explanation for the **gravitational lensing effect:** light, like all electromagnetic radiation, is deflected in a gravitational field. During the solar eclipse of 29 May 1919, the prediction of Einstein's theory that starlight is deflected by the solar mass by 1.75 arc seconds on its way to Earth was metrologically confirmed.
- According to the General Theory of Relativity, a sufficiently compact mass deforms space–time so strongly that a so-called **black hole** is formed in space. Neither matter nor information can escape from a black hole through electromagnetic radiation. Since the 1970s, the existence of black holes has been astronomically proven several times.
- According to Einstein's equations, the universe expands. From these equations, the **big-bang hypothesis** of the origin of the earth can also be assumed, according to which the origin of the expanding universe must lie in the explosion of an extremely compact and hot object. Saul Perlmutter and Brian P. Schmidt were awarded the 2011 Nobel Prize in Physics for their discovery of the accelerated expansion of the universe by observing distant supernovae.
- Albert Einstein postulated the existence of **gravitational waves** in 1916. A gravitational wave is a wave in space–time, caused by an accelerated mass. Since nothing can move faster than the speed of light, local changes in the gravitational field can only affect distant places after finite time. When passing through a space area, they temporarily compress and stretch distances within the space area. This can be seen as compression and stretching of space itself. The first direct measurement of gravitational waves—caused by the collision of two black holes—succeeded in September 2015 (Nobel Prize for Physics 2017, Rainer Weiss, Barry Barish, Kip Thorne).

The detection of gravitational waves was a scientific event of the century. This will not only be honored with a Nobel Prize in Physics, but also with a 70-cent stamp from the German postal service (Die Welt, November 21, 2017).

The special postage stamp "Gravitational waves" from the "Astrophysics" series is issued at December 2017. The stamp shows how gravitational waves are created when two black holes merge. This is a computer simulation carried out by scientists at the Max Planck Institute for Gravitational Physics in Potsdam.

3.2.6 Quantum Gravity: Theory of Quantization of Space and Time

Quantization is generally defined as the division of a whole into parts, so that certain quantities cannot take on any arbitrary value, but only fixed discrete values. In physics, quantization is known and experimentally proven for the following physical objects and quantities, among others: matter, momentum, angular momentum, light, electric charge, electric resistance. That the gravitational field has quantum properties is a generally shared conviction in physics, even if it has so far been supported solely by theoretical arguments and not by proof through experiments.

Quantum gravity is a theory to describe the structures of space and time in the smallest dimensions. The theoretical physicist Carlo Rovelli, who was involved in the development of this theory, describes the main features of the theory in his book *Reality that is not as it seems* (Rowohlt 2016) as follows:

- *The concept of loop quantum gravity represents an attempt to unite general relativity and quantum mechanics. The idea is simple. According to general relativity, space is the gravitational field. Quantum mechanics in turn teaches us that any such field is made up of quanta. From this it follows directly that space is made up of quanta of quantum fields. And time arises from the processes of these very fields. Space and time, as we perceive them on a large scale, are the blurred and approximate image of the gravitational field. The combination of general relativity and quantum mechanics points to something that is fundamentally different from our instinctive ideas about space, time and matter. According to the theory of loop quantum gravity, space exists from discrete volume pieces of the minimum size of a "Planck volume (10^{-99} cm^3) and time progresses in jumps of the order of magnitude of a "Planck time" (10^{-43} s).*

Rovelli shows the stages of development of the representation of space and time in the theory of quantum gravity in the diagram shown in Fig. 3.15.

3.3 Elementary Forces and the Origin of the Universe

The question of the "primordial forces of nature" has always occupied philosophers and natural scientists. As is well known, Goethe expresses it in his work Faust in

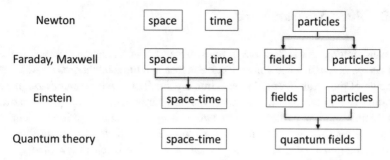

Fig. 3.15 The development phases of the theory of quantization of space and time

such a way ... *that I recognize what holds the world together in its innermost being* ...

According to the state of our knowledge, there are four elementary forces (elementary interactions), which underlie all physical phenomena of nature. Two of the elementary forces—the weak and the strong nuclear force—have ranges of 10^{-18} m and 10^{-15} m, respectively. They are therefore elementary forces that only have a subatomic effect and whose force effect can only be experimentally proven by "particle probes" (scattering experiments). The other two elementary forces—the electromagnetic force and the gravitational force—have an unlimited range and are valid throughout the universe. Because of their effects that reach into infinity, the elementary forces of many particles can be superimposed to form macroscopically measurable forces, and electromagnetic waves and gravitational waves can be used to generate energy and information into the whole universe.

3.3.1 Strong Nuclear Force

This elementary force—also called strong interaction or gluon force—is responsible for the extraordinarily large binding forces between the particles in the atomic nucleus (protons, neutrons). The range of the nuclear forces is of the order of the nuclear radius. Quantum chromo dynamics (QCD), developed since the 1960s, states that the smallest building blocks of matter, the quarks, are held together by "gluons" as exchange particles. The attraction between quarks increases with increasing distance (like a rubber band). The strong force is the basis of nuclear energy and thus also the cause of the sun's radiation energy.

3.3.2 Weak Nuclear Force

The weak interaction is decisive in transformations of elementary particles, e.g. in the β decay, in which a neutron emits an electron and turns into a proton. At 10^{-18} m, the range of the weak interaction is extremely small. The weak interaction plays an important role in the fusion reactions of solar energy, which is of vital importance to us. In an important partial reaction, two hydrogen nuclei fuse to form a deuterium nucleus by converting a proton into a neutron. The mathematical recording of the weak interaction was achieved in the 1960s with the electroweak theory, which also describes the electromagnetic force. The theory postulates that the weak force is transmitted by three exchange particles (*Z-boson*, *W⁺-boson*, *W⁻-boson*), which were generated at the European research center CERN in 1983, for which the Nobel Prize in Physics was awarded in 1984 (Carlo Rubbia, Simon van der Meer).

3.3.3 Electromagnetic Force

The electromagnetic force acts between electric charges. It can be shielded, eliminated (compensation of positive and negative charges) and is decisive for condensed matter, elementary processes in electrical engineering and electronics as well as for chemical and biological processes. The law of forces (Coulomb law) shows the distance dependence of the force $F \sim r^{-2}$ with infinite range. According to the quantum electrodynamics (QED) developed and experimentally confirmed in the 1940s, the force between electrically charged particles is transmitted by massless exchange particles, the photons or light particles.

Electromagnetic waves—these include radio waves, microwaves, light rays—consist of oscillating electrical and magnetic forces and propagate at the speed of light. An overview of the entire electromagnetic spectrum with wavelength in meter is given in Fig. 3.16.

Electromagnetic waves can transmit information over long distances. When they meet charged particles, such as the electrons in a radio or TV antenna, they

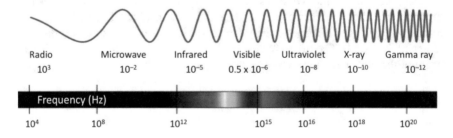

Radio	Microwave	Infrared	Visible	Ultraviolet	X-ray	Gamma ray
10^3	10^{-2}	10^{-5}	0.5×10^{-6}	10^{-8}	10^{-10}	10^{-12}

Frequency (Hz)

| 10^4 | 10^8 | 10^{12} | 10^{15} | 10^{16} | 10^{18} | 10^{20} |

Fig. 3.16 Illustration of the Electromagnetic Spectrum

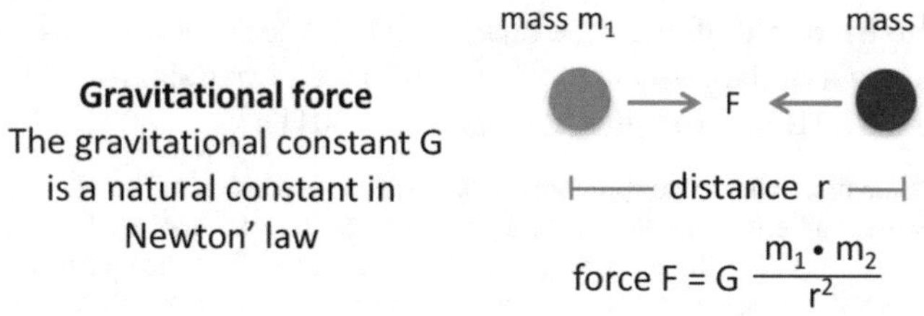

mass m_1 mass m_2

Gravitational force
The gravitational constant G
is a natural constant in
Newton' law

$\longrightarrow$ F $\longleftarrow$

$\longmapsto$ distance r $\longmapsto$

$$\text{force } F = G \, \frac{m_1 \bullet m_2}{r^2}$$

Fig. 3.17 Illustration of the Law of Gravity

excite the electrons and can discharge the wave information in them. After suitable amplification, the information can be used by humans as an audio or video signal sequence.

3.3.4 Gravitational Force

The gravitational force acts between masses and is 10^{-36} times weaker than the electromagnetic force. It cannot be shielded and manifests itself in the movement of the planets (Kepler's laws) and in the force of gravity (Newton's law of gravity). The law for the force F as a function of the range r shows, as in the case of the electromagnetic force, a proportionality of force and distance square ($F \sim r^{-2}$) and means an infinite range for the force effect, Fig. 3.17.

According to the general theory of relativity (see Sect. 3.2), gravitational waves consist of an oscillating spatial distortion, i.e. an oscillating stretching and compression of space. The local spatial distortion is inversely proportional to the distance from the source in space and thus extremely small on Earth. As already described in the previous section, the first direct detection of gravitational waves was achieved on September 14, 2015 by measuring the local space distortion caused by a gravitational wave using LASER interferometry (Advanced-LIGO detector) in the USA. The measured signal came from two black holes orbiting each other, getting closer and closer and finally merge The colliding objects combined between 7 and 36 solar masses. The gravitational wave signal was simultaneously detected by three detectors in the USA and Italy which significantly increased the accuracy of the experimental proof of gravitational waves.

3.3.5 Primordial Force

The overview of the four "elementary forces of nature", described in a short form, is today a secure physical knowledge. It is assumed that our universe was created about 14 billion years ago in a kind of "big bang" and that a single "primordial force" dominated the event. In the course of the expansion of the cosmos, one elementary force is said to have split off after the other until the forces known to us today were formed. The origin of the elementary forces from the primordial force is described as follows in the PTB publication *Maßstäbe* (May 2011):

> Gravity was the first to leave the compound as soon as the expansion of the universe began. 10^{-36} s after the Big Bang, when the temperature had cooled down to 10^{27} K, the strong force followed, and 10^{-12} s after the Big Bang, at a temperature of 10^{16} K, the electroweak force split into the weak and the electromagnetic force. Right after the Big Bang, the universe must have expanded dramatically. The driving force behind this accelerated expansion or "cosmic inflation" of the universe was gravity. However, the strong, weak and electromagnetic interactions determine which elementary particles appear in the universe and how they react with each other. Among other things, the nuclear building blocks, the proton and the neutron, were created in this process. It was so hot in the universe that the electrically charged atomic nuclei and electrons did not stick together, but formed an ionized, opaque gas. When the temperature had fallen so far that neutral atoms could form, the universe became lightning-fast transparent. This flash of light still fills the universe today as astrophysical verifiable "cosmic microwave background radiation" and is regarded as proof of the big bang theory.

3.3.6   The Origin of the Universe

The question of the origin of the universe cannot be answered in a simple way by present-day physics, since before the Big Bang the elementary reference parameters of physics—space and time—did not exist. Stephen Hawking expresses his thoughts about the origin of the universe in his posthumously published book *Short Answers to Big Questions* (Klett-Cotta, 2018) as follows:

> I think the universe was spontaneously created out of nothing, but completely in accordance with the laws of nature. Despite the complexity and diversity of the universe, it turns out that one only need three ingredients, namely "matter", "energy "and "space". Since we can imagine mass as a form of energy according to Einstein's famous formal $E = mc^2$, only two ingredients are needed to create a universe: "energy" and "space". After decades of research, cosmologists have found the answer: space and energy were spontaneously generated during an event we now call the Big Bang. At the moment of the big bang, time also began.

> The great puzzle of the Big Bang is the question of how a complete, huge universe full of space and energy can materialize from nothing. The mystery is explained by laws of physics that require the existence of a phenomenon we call "negative energy". When the Big Bang produced an enormous amount of positive energy, it simultaneously produced the same amount of negative energy. In this way, the positive and the negative always complement each other to zero. Where is all this negative energy today? It is in space. According to the laws of nature concerning gravity and motion, cosmic space itself is a huge store of negative energy. The infinite network of billions of galaxies, each of which acts with the force of gravity on all other galaxies, acts like a huge storage device.

> From astrophysical observations (redshift in the spectrum of moving objects) we now know that the universe is expanding. Based on the rate of expansion, we can estimate that about ten to 15 billion years ago, the galaxies must have been incredibly close together. Roger Penrose and I succeeded in proving geometric theorems that show that the universe was then completely confined to one point - a space–time singularity- must have been focused. At this point, Einstein's deterministic Theory of General Relativity loses its validity. In order to understand the origin of the universe, one must integrate the uncertainty relation of quantum mechanics in Einstein's theory of general relativity. Quantum–mechanical objects

can indeed appear completely random, stay for a while, only to disappear again and appear somewhere else. Since we know that the universe itself was once extremely small - smaller than a proton - the universe could have simply appeared out of nowhere without violating the laws of nature.

3.4 Measurement in Physics and Technology

Physics explores nature through observations, experiments, and measurements. The science of measurement is called *Metrology*. The International Vocabulary of Metrology (VIM) defines measurement as follows:

- *Measurement is the process of experimentally obtaining quantity values that can reasonably be attributed to a quantity. The quantity intended to be measured is called measurand.*

3.4.1 The International System of Units

In the middle of the nineteenth century, the need for a worldwide decimal metric system became very apparent as a consequence of the Industrial Revolution. In 1875, a diplomatic conference took place in Paris, where 17 governments signed the diplomatic treaty, the "Metre Convention". The signatories decided to create and finance a permanent scientific institute: the "Bureau International des Poids et Mesures" BIPM, www.bipm.org. As of January 2020, there are 62 member states and 41 associated states and economies.

In 1946, the MKSA system (metre, kilogram, second, ampere) was accepted by the Metre Convention countries. The MKSA was extended in 1954 to include the kelvin (temperature measurements) and candela (optical measurements). The system then obtained the name the *International System of Units, SI* (*Le Système International d'Unités*). Representatives of the governments of the member states meet every fourth year for the "Conférence Générale des Poids et Mesures" (CGPM). At the 14th CGPM in 1971, the SI was again extended by the addition of the *mole* as base unit for amount of substance.

Following the redefinition of the meter in 1983, which set the speed of light at 299,792,458 m/s as natural constant, there has been considerable discussion on the redefinitions of a number of the base units. The solution to this problem is to avoid "material properties" to define a unit, but to use "dematerialized natural constants ". The new SI system is comprised of seven base units. They are named in the following together with the uncertainty of their realization by Reference Standards, stated by the Committee on Data for Science and Technology, CODATA (www.codata.org):

- *Time*: second (s). The duration of 9,192,631,770 periods of the radiation corresponding to the transition between the two hyperfine levels of the ground state of the caesium -133 atom [2×10^{-16}].

- *Length*: meter (m). The distance travelled by light in vacuum in 1/299,792,458 s $[10^{-12}]$.
- *Mass*: kilogram (kg). The kilogram is defined by setting the Planck constant h exactly to $6.62607015 \times 10^{-34}$ J·s ($J = kg·m^2·s^{-2}$), given the definitions of the meter and the second $[2 \times 10^{-8}]$.
- *Electric current*: ampere (A). The flow of $1/1.602176634 \times 10^{-19}$ times the elementary charge e per second $[9 \times 10^{-8}]$.
- *Temperature*: kelvin (K), The kelvin is defined by setting the fixed numerical value of the Boltzmann constant k to 1.380649×10^{-23} J·K^{-1}, ($J = kg·m^2·s^{-2}$), given the definition of the kilogram, the meter, and the second $[3 \times 10^{-7}]$.
- *Amiunt of substance*: mol (N). The amount of substance of exactly $6.02214076 \times 10^{23}$ elementary entities. This number is the fixed numerical value of the Avogadro constant N_A, when expressed in the unit mol $[2 \cdot 10^{-8}]$
- *Luminous intensity*: candela (cd). The luminous intensity, in a given direction, of a source that emits monochromatic radiation of frequency 5.4×10^{14} Hz and that has a radiant intensity in that direction of 1/ 683 W per steradian $[10^{-4}]$.

An overview of the base units of the International System of Units together with the natural constants on which they are based is given in Fig. 3.18.

The introduction of the new International System of Units was highlighted by the German Physical Society (DPG) under the heading "Paradigm Shift in Physics" as follows:

- **Natural Constants as the Measure of all Things**
  On May 20, 2019 (World Metrology Day), the International System of Units will undergo a fundamental change in accordance with the resolution of the General Conference of the Meter Convention of November 2018: Natural constants will then define all physical units, and the units of measurement of technology based

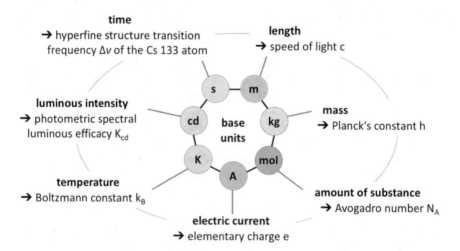

Fig. 3.18 The new International System of Units and its fundamentals

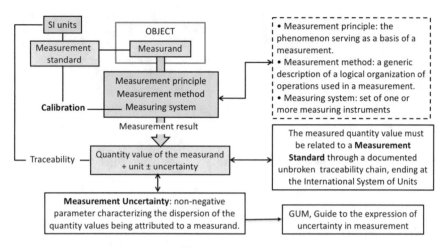

Fig. 3.19 The methodology of measurement

on them. Thus, the SI is open to all technological innovations for all times and at any place in the world (DPG Publication PHYSIKonkret, September 2018).

3.4.2   Methodology of Measurement

Measurement begins with the definition of the measurand, the quantity intended to be measured. When the measurand is defined, it must be related to a measurement standard, the realization of the definition of the quantity to be measured. The measurement procedure is a detailed description of a measurement, according to a measurement principle and to a given measurement method. It is based on a measurement model and including any calculation to obtain a measurement result. The result of a measurement has to be expressed as a quantity value together with its uncertainty, including the unit of the measurand. The methodology of measurement is illustrated in Fig. 3.19.

The operational procedure to perform a measurement is illustrated in Fig. 3.20 with the example of the measurement of temperature.

The measured quantity value must be related to a Measurement Standard through a documented unbroken "traceability chain", Fig. 3.21.

3.4.3   The Metrological Concept of Measurement Uncertainty

The International Vocabulary of Metrology (VIM) defines *measurement uncertainty* as a "non-negative parameter, characterizing the dispersion of the quantity values being attributed to a measurand". The master document, which is acknowledged

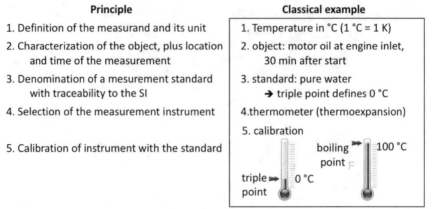

Principle	Classical example
1. Definition of the measurand and its unit	1. Temperature in °C (1 °C = 1 K)
2. Characterization of the object, plus location and time of the measurement	2. object: motor oil at engine inlet, 30 min after start
3. Denomination of a mesurement standard with traceability to the SI	3. standard: pure water → triple point defines 0 °C
4. Selection of the measurement instrument	4. thermometer (thermoexpansion)
5. Calibration of instrument with the standard	5. calibration

6. Measurement: determination of quantity values of the measurand → mean value

7. Determination of measurement uncertainty → standard deviation

> Result of measurement:
> Quantity value of the measuran + unit ± uncertainty

Fig. 3.20 The operational principle of performing a measurement and a classical example

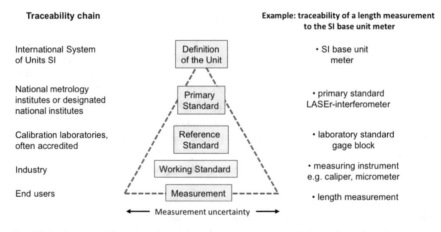

Fig. 3.21 The traceability chain for measurements and an example from dimensional metrology

to apply to all measurement and testing fields and to all types of uncertainties of quantitative results, is the *Guide to the Expression of Uncertainty in Measurement* (GUM). The essential points of the "GUM uncertainty philosophy" are:

– A measurement quantity, of which the true value is not known exactly, is considered as a stochastic variable with a probability function. Often it is assumed that this is a normal (Gaussian) distribution.

- measured quantity values x_i: x_1, x_2, ...,
- arithmetic mean $\quad \bar{x} = \dfrac{1}{n}\sum_{i=1}^{n} x_i$
- standard deviation $\quad s = \sqrt{\dfrac{1}{n-1}\sum_{i=1}^{n}(x_i - \bar{x})^2}$
- standard measurement uncertainty $u = s$
- expanded measurement uncertainty: $U = k \cdot u$
- result of measurement: $\bar{x} \pm u_x$

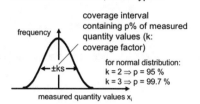

Statistical evaluation, GUM type A

frequency

coverage interval containing p% of measured quantity values (k: coverage factor)

for normal distribution:
$k = 2 \Rightarrow p = 95\%$
$k = 3 \Rightarrow p = 99.7\%$

$\pm ks$

measured quantity values x_i

Fig. 3.22 The evaluation of uncertainty of measurements

- The result x of a measurement is an estimate of the expectation value.
- Expectation (quantity value) and standard uncertainty are estimated either by statistical processing of repeated measurements (Type A, Uncertainty Evaluation) or by other methods (Type B, Uncertainty Evaluation).
- The result of a measurement has to be expressed as a quantity value together with its uncertainty, including the unit of the measurand.

The methodology of the determination of metrological uncertainty, GUM Type A, has the following steps, Fig. 3.22.

- Identify the important components of measurement uncertainty. Apply a model of the actual measurement process to identify the sources, if possible.
- Calculate the standard uncertainty of each component of measurement uncertainty: Each component of measurement uncertainty is expressed in terms of the standard uncertainty determined from either a Type A or Type B evaluation.
- Calculate the expanded uncertainty, U: Multiply the combined uncertainty with the coverage factor k ($U = k\,u$).
- State the measurement result in the form $X = x \pm U$, with coverage factor k. For a normal distribution and a coverage factor $k = 3$, the coverage interval ± 3 s (mathematically $\mu \pm 3\sigma$) contains 99.7% of the measured quantity values.

3.4.4 Multiple Measurement Uncertainty

The method outlined in the foregoing section considers only one single measurement quantity. However, very often uncertainty evaluations have to be related to functional combinations of measured quantities or uncertainty components $y = f(x_1, x_2, x_3..., x_n)$. In these cases, for uncorrelated (i.e. independent) values, the single uncertainties are combined applying the law of propagation of uncertainty to give the so called *combined measurement uncertainty* $u_{combined}(y) = \sqrt{\sum (\partial f/\partial x_i)^2 u^2(x_i)}$, where $u(x_i)$ is the uncertainty of the variable x_i.

From the statistical law of the propagation of uncertainties it follows that there are three basic relations, for which the resulting derivation becomes quite simple:

(a)   for equations of the measurand involving only sums or differences:

$$y = x_1 + x_2 + \ldots + x_n, \text{ it follows } u_y = \sqrt{\left(u_3^2 + u_3^2 + \ldots + u_n^2\right)}$$

(b)   for equations of the measurand involving only products or quotients:

$$y = x_1 \cdot x_2 \cdot \ldots \cdot x_n \text{ it follows } u_y/|y| = \sqrt{\left(u_1^2/x_1^2 + u_2^2/x_2^2 + \ldots + u_n^2/x_n^2\right)}$$

Example: The uncertainty u_P of the quantity value of mechanical power, $P =$ force $F \cdot$ velocity v, where F and v have relative uncertainties u_F and u_v of 1%, is given by $u_P = \sqrt{(1^2 + 1^2)} = 1{,}4\%$.

(c)   for equations of the measurand involving exponents

$$y = x_1^a \cdot x_2^b \cdot \ldots \cdot x_n^2 \text{ it follows } u_y/|y| = \sqrt{\left(a^2 u_1^2/x_3^2 + b^2 u_2^2/x_2^2 + \ldots + z^2 u_n^2/x^2\right)}.$$

3.4.5   Accuracy of Measurements

Accuracy is an umbrella term characterizing the closeness of agreement between a measurement result and a reference value of a measurand. If several measurement results are available for the same measurand from a series of measurements, accuracy can be split up into *trueness* and *precision*. The terms trueness and precision are defined in the International Standard ISO 3534.

– *Trueness* accounts for the closeness of agreement between the mean value x_m and the true (reference) value. The difference between the mean value x_m and the true (reference) value is often referred to as *systematic error* S.
– *Precision* describes the closeness of agreement Δ of the individual values themselves. A small value of Δ indicates high precision.

The *target model*, illustrated in Fig. 3.23 visualizes the different possible combinations, which result from true or wrong and precise or imprecise results.

3.4.6   Measuring Instruments: Calibration and Accuracy Classes

All measurements must be performed with appropriate *measuring instruments*. A measuring instrument gives as output an "indication", which has to be related by **calibration** to the quantity value of the measurand. A calibration diagram represents the relation between indications of a measuring instrument and a set of reference values of the measurand. **Accuracy classes of measuring instruments** are denoted by a number or symbol as follows:

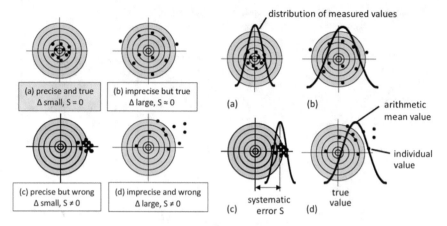

Fig. 3.23 The target model to illustrate precision and trueness of measurements

- In analog instruments, an accuracy class of 1.0 indicates that the limits of error—in both directions—should not exceed 1% of the full-scale deflection.
- In digital instruments, the limit of indication error is ±1 digit of the least significant unit of the digital indication display. For example, the maximum relative permissible measurement error for two decades is ± 1/99 ≈ 1%, and for three decades it is 1/999 ≈ 0,1%.

An example of calibration from dimensional metrology and the definition of the measuring instrument accuracy class is given in Fig. 3.24.

3.4.7 Sensors

Sensors are measurement transducers. They transform parameters of interest into electrical signals that can be displayed, stored or further processed for applications in science and technology.

3.4.7.1 Physics of Sensors

There is a broad variety of physical principles that can be used to determine and measure parameters of interest in physics and technology. The following brief compilation names a few examples.

- *Bragg's law*: An optical light beam diffracted by a crystal lattice is related to the crystal plane separation and the wavelength of the beam.
  - This effect is used to measure the crystal lattice geometry and stress-induced strain of a crystalline specimen with a Fiber Bragg Grating Sensor.

(1) Calibration of measuring instrument (measurand: length)

Reference standard: gauge block (traceable to the Si length unit with an optical interferometer)

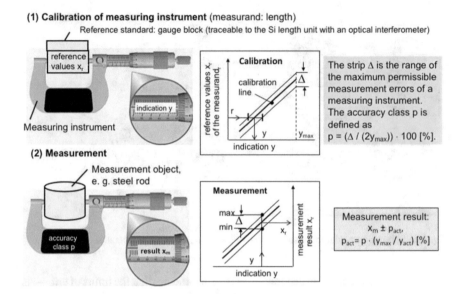

Fig. 3.24 Calibration example from dimensional metrology

- *Poisson effect*: A material deforms in a direction perpendicular to an applied stress.

  – This effect is partially responsible for the response of a strain gage.

- *Raoult's effect*: Resistance of a conductor changes when its length is changed.

  – This effect is partially responsible for the response of a strain gage.

- *Snell's law*: Reflected and refracted rays of light at an optical interface are related to the angle of incidence.

  – Fibre-optic strain sensors are based on this law.

- *Villari effect*: The magnetoelastic effect of the change of magnetic susceptibility of a ferromagnetic material when subjected to a mechanical stress.

  – Magnetostrictive position sensors are based on this effect.

Physical sensor effects for kinematics

- *Bernoulli's equation*: Conservation of energy in a fluid predicts a relationship between pressure and velocity of the fluid.

  – A Pitot tube uses this effect to measure air speed of an aircraft.

- *Coriolis effect*: A body moving relative to a rotating frame of reference experiences a force relative to the frame.

- A coriolis gyrometer detects disturbing torque moments acting detrimental on a moving automobile.

- *Doppler effect*: The frequency received from a wave source (e.g., sound or light) depends on the speed of the source.

  - A laser doppler velocimeter uses the frequency shift of laser light reflected off of moving bodies, e.g. machinery components or moving automobiles.

- *Gauss effect*: The resistance of a conductor increases when magnetized.

  - This effect is used to determine lateral or rotational motions of moving components in machines or in automobiles.

  - *Hall effect*: A voltage is generated perpendicular to current flow in a magnetic field.

    - A Hall effect proximity sensor detects when a magnetic field changes due to the motion of a metallic object.

Physical sensor effects for dynamics

- *Lorentz's law:* There is a force on a charged particle moving in an electric and magnetic field.

  - The Lorentz force is the basic effect for the operation of motors and generators.

- *Newton's law:* Acceleration of an object is proportional to force acting on the object.

  - This law is essential for acceleration sensors.

- *Piezoelectric effect:* Charge is displaced across a crystal when it is strained.

  - A piezoelectric accelerometer measures charge polarization across a piezoelectric crystal subject to deformations due to a force.

Physical sensor effects for temperature

- *Biot's law:* The rate of heat conduction through a medium is directly proportional to the temperature difference across the medium.

  - This principle is basic to time constants associated with temperature transducers.

- *Joule's law:* Heat is produced by current flowing through a resistor.

  - The design of a hot-wire anemometer is based on this principle.

- *Seebeck effect:* Dissimilar metals in contact result in a voltage difference across the junction that depends on temperature.

  - Principle of a thermocouple.

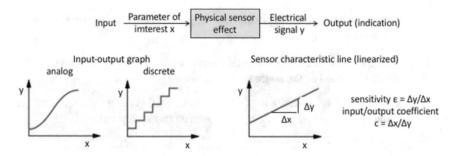

Fig. 3.25 Block diagram of a sensor and the graphical representation of its function

- *Stefan-Boltzmann law:* The heat radiated from a black body is proportional to the fourth power of its temperature.

  – Principle for the design of a temperature-sensing pyrometer.

3.4.7.2   Sensor Characteristics and Sensor Types

A sensor can be illustrated by a block diagram and described by the sensor function $y = f(x)$ which relates the output y to the input x, see Fig. 3.25 If there is a linear input–output relation, the sensitivity ε of the sensor is defined as $\varepsilon = \Delta y / \Delta x$, and its reciprocal is the input/output coefficient $c = \Delta x / \Delta y$. From the measurement of y, the parameter of interest x can be determined through the relation $x = c \cdot y$.

A broad variety of sensors have been developed for the determination of parameters of interest in physics and technology, which can be classified according to their input–output relations. An overview of sensor categories is given in Fig. 3.26. Fig. with Fig. 3.26a a sensor selection matrix and Fig. 3.26b the principles of wave propagation sensors.

Wave propagation sensors for the determination of distance and speed operate with transmitted and reflected waves of different wavelength ranges, categorized as follows:

RADAR (Radio detection and ranging)

RADAR is an object-detection system that uses radio waves to determine the range, angle, or velocity of objects. A radar system consists of a transmitter producing electromagnetic waves in the radio or microwave domain, a transmitting antenna a receiving antenna (often the same antenna is used for transmitting and receiving) and a receiver and processor to determine properties of the object. Radio waves (pulsed or continuous) from the transmitter reflect off the object and return to the receiver, giving information about the object's location. The velocity v of a moving object, e. g. the speed of a car, can be detected with the Doppler shift.

LIDAR (Light detection and ranging)

(a)

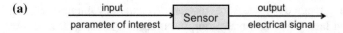

Output / Input	Electrical signal				
	resistance	inductance	capacitance	voltage	current
strain $\varepsilon = \Delta l/l$	strain gage				fiber optical sensor
position: length l, distance s angle	resistive sensor, magneto-resistive sensor	inductive position sensor	capacitive position sensor	Hall sensor, triangulation sensor	eddy current sensor, opto-electric sensor
velocity $v = ds/dt$	magneto-resistive angle sensor	inductive angle sensor		magnetopole sensor	optoelectronic revolution sensor
acceleration $a = dv/dt = d^2s/dt^2$	seismic sensors with dynamic mass-damper-spring elements, sensing of mass displacement with electronic position sensor (e. g. Hall sensor)				
force F **torque F• l**	piezo-resitive sensor, strain gauge sensor	magneto-elastic sensor		piezoelectric sensor	sensors with elastic elements and strain or displacement sensor
pressure p = force/area	piezo-resitive sensor				
temperature T	NTC/ PTC resistor			thermo-couple	optoelectronic pyrometer

(b)

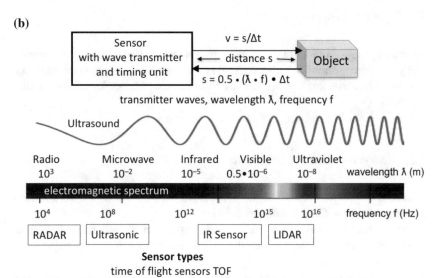

Fig. 3.26 a Overview of sensor types and a sensor selection matrix. **b** The physics of wave propagation for sensing distance and speed

LIDAR is a surveying method that measures distance to a target by illuminating that target with a pulsed laser light, and measuring the reflected pulses with a sensor. Differences in laser return times and wavelengths can then be used to measure distances. LIDAR uses Ultraviolet, Visible, or near Infrared light to image objects. It can target a wide range of materials, including non-metallic objects, rocks, and chemical compounds. A narrow laser-beam can map physical features with very high resolution.

IR Sensor

An IR sensor (as sensor array called an infrared camera or thermal imaging camera) is a device that forms an image using infrared radiation up to $\lambda = 14\,\mu m$. This sensor technique can also be used for remote-sensing temperature.

Ultrasonic Sensor

In a similar way to RADAR, ultrasonic sensors evaluate targets by interpreting the reflected signals. For example, by measuring the time between sending a signal and receiving an echo, the distance of an object can be calculated.

3.5 Physical Observations

Since the beginning of modern times, numerous natural scientists have made important physical observations, which are listed below with keywords of selected examples in roughly chronological order. They form building blocks for the world view of physics.

3.5.1 Astronomical Phenomena

- Nicolaus Copernicus describes the heliocentric world view in his work *On the Revolutions of the Celestial Bodies* (1541).
- Through astronomical observations of a supernova in the constellation of Cassiopeia, Tycho de Brahe refutes the Aristotelian assumption of the unchanging celestial sphere (1572).
- Based on the precise celestial observations of Tycho de Brahe, Johannes Kepler formulated the mathematical law that planets move on elliptical orbits with the Sun at one focal point (1605).
- Earth's atmosphere and vacuum are discovered by Otto v. Guericke and the effect of air pressure is demonstrated with the (airless) Magdeburg hemispheres (1654).
- The rotation of the earth is proved by Foucault with a pendulum (28 kg mass, 67 m length, Pantheon in Paris) (1851).
- George Gamow develops the theory of the *Big Bang* as the standard model of cosmology to explain the beginning of the universe (1948).

3.5.2 Optical Phenomena

– Snellius formulates the law of optical refraction: Change of direction of a light beam during transition to another medium (1620).
– Pierre de Fermat established the *extrememal principle of optics* named after him, from which the law of reflection and the law of refraction can be derived (1657).
– Isaac Newton demonstrates the spectral decomposition of sunlight, develops colour theory and establishes the corpuscular theory of light (1672).
– Christian Huygens develops the wave theory of light; according to the principle named after him, every point on a wave front can be considered the starting point of a new wave (1673).
– Thomas Young describes the superposition of wave trains (interference) and develops the three-color theory of color perception (1802).
– The speed of light is described by Fizeau with a value of 298,000 km/s (gear wheel method 9 m measuring distance) (1848).
– Albert Einstein explains the "origin of light" with the light quantum hypothesis by the emission of photons (light quanta) as a result of quantum leaps in atoms (1905) and postulated in 1916 the stimulated emission of light as the theoretical basis for the LASERs (Light Amplification by Stimulated Emission of Radiation) realized from 1960 onwards.

3.5.3 Mechanical Phenomena

– Galilei performs motion experiments on the inclined plane, discovers the laws of gravity of mechanics and founded the theory of elasticity by deformation experiments on a beam (main work Discorsi 1638).
– Isaac Newton established the law of gravity (1666) and formulated the classical Mechanics with the famous three Newtonian axioms.
– The gravitational constant is determined by Cavendish using a torsion balance (1798).
– According to Robert Hooke, the elasticity of solids is determined by a linear relationship between force and spring extension (1678).
– According to Thomas Boyle, the general gas equation (ideal gas) applies to gases in the formula pressure x volume = amount of substance x gas constant x temperature (1662).
– For the mechanics of fluids (gases and liquids), Bernoulli discovered the relationship between flow velocity and pressure (1738).
– Chladni determines the speed of sound in liquids and solids (1796).
– Mach investigates the supersonic bang that occurs at flight speeds greater than the speed of sound (340.3 m/s) (1887).
– Einstein extended the classical view of space and time with the special theory of relativity (1905).

3.5.4   Thermal Phenomena

– To determine the temperature, Galilei invented the thermoscope, a preform of the
  thermometer (1592). Celsius proposes a 100-part thermometer division (1742).
– Benjamin Thompson, Earl of Rumford determines the mechanical heat equivalent
  and recognizes the non-materiality of heat: heat is a form of movement of the
  smallest particles (1798).
– The law of conservation of energy for heat processes (1st law of thermodynamics) is formulated by Mayer and confirmed by Joule and Helmholtz (1842):
  the total energy remains constant during conversion processes thermal energy
  ↔mechanical energy.
– The 2nd law of thermodynamics is established by Clausius: Heat always passes
  by itself only from a warmer body to a colder body, never the other way around.
  Conclusion from the 1st and 2nd law: a *perpetuum motion machine* is impossible
  (1850).
– Sadi Carnot founded thermodynamics on observations of the steam engine;
  Carnot's cycle shows the greatest thermal efficiency (1824).

3.5.5   Electromagnetic Phenomena

– Basis of electrostatics is the law named after Coulomb about forces between
  electric charges (1785).
– The first electrochemical voltage source is realized by Volta (1799).
– Ørsted discovers the magnetic effect of electric current (electromagnetism) and
  invents the measuring device for measuring electrical current (1820).
– Ampere establishes a theory of the interactions of current-carrying conductors
  and explains magnetism by molecular currents (1821).
– Becquerel discovers piezoelectricity, the appearance of electric charges when a
  force is applied to certain crystals (direct piezoelectric effect), the indirect piezo-
  electric effect is the change in shape when a high electric voltage is applied
  (1819).
– Seebeck discovers thermoelectricity, the reversible interaction of temperature and
  electricity (1821).
– Ohm formulated the law named after him:
– Electrical voltage U = electrical resistance R · electrical current I (1826).
– Faraday discovers electromagnetic induction and magnetic field lines (1831).
– Joule measures the thermal effect of electric current (1842).
– Kirchhoff formulates Kirchhoff's rules for describing the relationship between
  several electric currents and between several electric voltages in electric networks
  (1845).
– Electromagnetism can be described by the equations formulated by Maxwell in
  mathematical form, The Maxwell equations describe the relationship of electric

and magnetic fields with electric charges and electric current under given boundary conditions (1865).

- The Lorentz force is the force exerted by a magnetic or electric field on a moving electric charge; it is the basis of electrodynamics (1895).
- Heinrich Hertz proves that radio waves move at the speed of light and justifies the representation of the totality of electromagnetic waves (radio waves, light, X-rays) as an electromagnetic spectrum (1894).
- Drude proves in the electron theory of metals that electricity in metals is based on directionally moving electrons (1900).
- The transistor, the fundamental component of electronics for switching and amplifying electrical signals, is developed by Bardeen, Brattain and Shockley (1948).

3.5.6 Quantum Physical Phenomena

- Max Planck established quantum theory to describe the physical phenomena of atomic physics, solid state physics and nuclear and elementary particle physics (1900).
- Nils Bohr develops Bohr's atomic model, which contains elements of quantum mechanics (1913).
- The quantization of energy in the atomic shell is experimentally confirmed by the Franck-Hertz experiment (1913).
- Louis de Broglie postulated the experimentally proven wave-particle dualism: Objects from the quantum world can be described both as particles (corpuscles) as well as a wave (1924).
- Wolfgang Pauli formulated the Pauli principle, which is fundamental to the construction of atomic shells (1925).
- For the representation of quantum theory in mathematical form, Werner Heisenberg developed matrix mechanics (1926) and Erwin Schrödinger wave mechanics (1927). John von Neumann proves the mathematical equivalence of both theories (1944).
- Heisenberg formulates the uncertainty relation as a statement of quantum physics that two measurable variables (observables) of a particle (e.g. location and momentum) cannot at the same time be determined as precisely as desired (1927).
- The basic model of nuclear physics is formulated: Atomic nuclei consist of protons and neutrons (Heisenberg 1932).
- Maria Göppert-Mayer develops the shell-model of the atomic nucleus (1949).
- The first nuclear fission is carried out by Hahn and Strassmann (1938).
- Nuclear fusion, the fusion of atomic nuclei with the release of energy as the cause of the sun's radiation energy is explained by the Bethe-Weizsäcker cycle (carbon–nitrogen cycle) (1938).

- Robert Hofstadter discovers by means of the scattering of high-energy electrons at light atomic nuclei the existence of internal structures in proton and neutron (1961).
- The quark model to describe the internal of the atomic nucleus is postulated by Gell-Mann (1963) and experimentally confirmed by scattering electrons on protons and neutrons (1970).

3.6 The Evolution of Physics

The classical fields of physics—*mechanics, heat* (*thermodynamics*) and *optics*—emerged from a broad variety of physical observations, their experimental verification and mathematical representation. The field of *electromagnetism* developed from the observation of electrical and magnetic phenomena and the investigation of their interrelationships. The entirety of optical and electromagnetic radiation processes can be described by the *electromagnetic spectrum*. The named areas constitute Classical Physics which is complete in its scope and experimentally confirmed.

With the new space–time understanding of the theory of relativity (1905) and quantum theory (1920s) for the description of physical phenomena of the micro- and nano-world, classical physics was extended to the current world view of physics.

It should be noted that in physics—in contrast to philosophy, where, according to the physicist and philosopher Carl Friedrich von Weizsäcker, *"each of the great philosophers understood the world in a way that was unique to him"*– the structure of physics follows the "correspondence principle". It states that an older scientific theory is contained in a newer theory, but the newer theory has an extended scope of validity (Ehrenfest theorem.) This can be illustrated by a comparison of *classical mechanics* and *quantum physics*. Quantum physics is not deterministic in the classical sense and operates with "probability predictions" for the value of a measured parameter such as the location of microscopic objects. If the rules of quantum physics are applied to macroscopic mechanical systems, the statistical dispersion of the measurement results becomes almost immeasurably small and the mean values of quantum mechanical quantities are transformed into the classical equations of motion. Thus, the Ehrenfest theorem has proven that the quantum mechanical expectation values obey Newton's classical equations of motion. This means that both theories are true and complementary in their scope.

3.6.1 *Mechanics*

Mechanics is mainly concerned with movement processes (Kinematics) and the effects of forces (Dynamics). Depending on the object, a distinction is made between stereo mechanics (mass points, rigid bodies), continuum mechanics of fluids, and structural mechanics.

Kinematics considers the movement of mass points or bodies using the terms position, velocity (distance per unit of time) and acceleration (velocity per unit of time). Basic rules of kinematics:

1. A geometrical reference system is required to describe movements.
2. Movements consist of linear movements (translations) and rotary movements (rotations).
3. A spatially freely moving body has three degrees of freedom of translation and three degrees of freedom of rotational movement.

From the experimentally observed fact, that by "symmetry operations", namely (a) translation in space, (b) translation in time, kinematic laws do not change, conservation laws are derived: from (a) follows the law of conservation of momentum and from (b) follows the law of conservation of energy.

Dynamics deals with forces in a state of equilibrium (statics), the relationship between forces and motion (kinetics) and the influence of forces on masses and bodies (structural dynamics). The basic rules of dynamics (Newton's axioms) are.

1. Principle of inertia: A body remains at rest or in uniform motion unless it is forced to change its state by a force.
2. Principle of action (force-movement relation) expressed in the formulation of Leonard Euler (1750): force = mass $\cdot$ acceleration.
3. Reaction principle: forces occur only in pairs, (actio equals reactio)

Further basic concepts of dynamics are:

- momentum = mass $\cdot$ velocity;
- work (mechanical energy) = force $-$ speed;
- power = work/time.

Structural Mechanics is a field of applied mechanics that investigates the behavior of structures under mechanical loading. A simplified model illustration is given in in Fig. 3.27. *Stress*, $\sigma = F/A_0$, gives the intensity of a mechanical force F passes through a body's cross-sectional area, A_0. S*train*, $\varepsilon = \Delta l/l_0$, gives the relative displacement of the body. Stress–strain curves depend on materials composition and microstructure, they are influenced by the loading manner, strain rate, temperature and chemical environment. From the experimental determination of stress–strain curves, three basic features of structural mechanics can be observed:

- Elasticity. When the applied load is small, reversible elastic deformation occurs and stresses are proportional to elastic strains (Hooke's law: $\sigma = E \cdot \varepsilon$). The slope between tensile stress and tensile elastic strain is called the Elastic (or Young's) modulus, E.
- Plasticity. When the applied stress becomes higher, materials (especially metals) usually show plastic deformation as a result of sub-microscopic dislocation movements. The stress at the onset of plastic flow is called the *yield strength,* which has no direct relation with the Young's modulus. The yield strength indicates

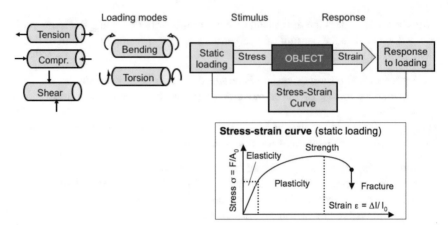

Fig. 3.27 Modes of loading in structural mechanics and the typical stress–strain curve of ductile materials for quasi-static tensile loading

resistance against plastic flow, and the ultimate tensile strength is an important mechanical property.

- Fracture. Under dynamic loading, plastic flow may occur. The "strength" of materials decreases with the number of load cycles, and final fracture takes place. The stress at 10^7 cycles is called the fatigue strength or endurance limit. Fatigue strength and fatigue life are utilized for the design of machines under dynamical loading.

3.6.2 Relativity

The special theory of relativity (Einstein 1904) describes kinematics from the point of view of observers in reference systems (coordinate systems) that move uniformly relative to each other. (The general theory of relativity (Einstein 1916) considers accelerated motion systems and attributes gravity to a "curvature" of space and time, which is caused by the masses involved see Sect. 3.2).

The Relativity theory is based on two experimentally verified observations:

1. The mass of a body m increases with increasing velocity v. It holds: $m = m_0/\sqrt{(1 - v^2/c^2)}$, where m_0 is the rest mass and c is the speed of light. The formula was confirmed with particle experiments near the speed of light. In movements of macroscopic bodies on earth, the difference between m and m_0 is practically unnoticeable Even at the very high satellite speed of 8 km/sec, the mass changes according to this formula is only a few billionths of the rest mass.
2. it was experimentally established that the speed of light has a constant value of $c \approx 300,000$ km/s for both stationary (resting) and dynamic (moving) senders (Michelson-Morley Experiment 1887).

The coordinate system **X–Y–Z** is stationary and the coordinate system **X–Y–Z** moves with the speed v_y

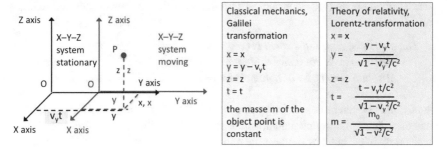

Fig. 3.28 Kinematics of an object point in classical mechanics and in the theory of relativity

The problem of motion in classical mechanics or relativity is solved by introducing coordinate systems and observing the trajectory of an object point. In Fig. 3.28, an object point P is shown in two coordinate systems moving relative to each other; x, y, z are the spatial coordinates and t is the time. At very high speeds, it follows, from the Lorentz transformation and the formulae for the parameters y and t, that length scales appear shortened (length contraction, also called Lorentz contraction) and moving clocks go slower (time dilation).

Another consequence of the principle of relativity is an equivalence of mass and energy, described by Heisenberg as follows: *Experimentally, one can see directly how elementary particles are produced from kinetic energy and how such particles can disappear again by transforming into radiation.*

At speeds near light speed c, the equivalence of mass m and energy E is expressed by the Einstein equation $E = m \cdot c^2$. If the associated velocities are less than the speed of light, the dependencies on the state of motion fall below the experimental detection limit. With sufficient accuracy, classical mechanics is then sufficient to describe the "physics of our sensory experience".

3.6.3 Heat

The theory of heat developed from the insight that heat is not a substance but the energy of the movement of smallest particles (impact processes in gases, lattice vibrations in solids) and that heat can be described as *thermodynamics* with methods of statistical mechanics. Temperature T (s, t) is a state variable dependent on location s and time t. Temperature is the macroscopic measure for the microscopic kinetic energy of the particles of a thermodynamic system in thermal equilibrium. The heat capacity is the ratio between the amount of heat supplied and the increase in temperature; it is a temperature-dependent material property. The thermal conductivity is a measure of the speed of heat transfer between two materials at a given temperature difference.

The basic principles of thermodynamics are derived from the observation of thermal phenomena (see Sect. 3.6). The fundamental theorems of thermodynamics are:

1. *The energy of a closed system is constant.*
2. *Thermal energy is not convertible into other types of energy to any degree.*
3. *The absolute zero point of temperature (0 K = −273,15 °C) is unattainable.*

The second law can be expressed by introducing the term entropy as follows: Entropy change = heat quantity change/absolute temperature. (In the model conception of thermodynamics, entropy denotes the number of microstates by which the observed macrostate of the system can be realized). Together with temperature, pressure and volume, entropy describes the state of a thermodynamic system.

3.6.4 Optics

Optics is the branch of physics that studies the behavior and properties of light including its interactions with matter and the construction of instruments that use or detect it. Light can be perceived by the human eye as optical radiation in the wavelength range from 380 nm (blue) to 780 nm (red). The light propagation can be described according to the experimentally confirmed *wave-particle dualism* as *wave optics* or as geometric optics.

- *Wave optics* treats light as an electromagnetic wave, which results in properties like color, interference ability; diffraction of light.
- *Geometric optics* assumes simplifying that the photons move on straight paths (rays). By applying the optical laws of reflection and refraction, by means of optical components (e.g. mirrors, lenses, prisms) the classical optical devices (e.g. microscope, telescope). Are designed

Light can be created by the following physical mechanisms, Fig. 3.29:

- in solids by radiation emission (incandescent lamp),
- in gases by shock excitation of atoms, ions, molecules (fluorescent lamp),
- in semiconductor structures by recombination of charge carriers (Light-emitting diode LED).
- LASER radiation is monochromatic, interference-capable (coherent) narrow bundled and very intense. Light pulses of 10^{-15} s duration (femtosecond LASER) and power densities of more than 1000 W/cm^2 with temperatures of several 1000 °C, and material evaporation can be realized.

A black body is an ideal source which emits thermal radiation, which is electromagnetic radiation of any wavelength. Absorption capacity and emissivities are proportional to each other (Kirchhoff's law of radiation). Planck's radiation law describes

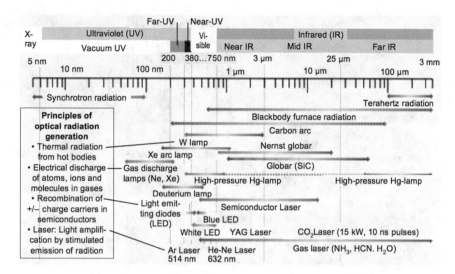

Fig. 3.29   Overview on optical light sources in regimes of the electromagnetic spectrum

the distribution of radiation energy in dependence on the wavelength of the radiation. An illustration of Planck's radiation law and its application in thermography is given in Fig. 3.30.

Light consists of photons (light quanta, see Fig. 3.6). The photon energy is called radiometric energy—and when evaluated according to the spectral luminous sensitivity of the human eye—is called photometric quantity of light. The radiation power (radiant flux) is the energy emitted in a time interval (photometric luminous flux). The radiance (photometric luminance) is the radiation flux per solid angle and per effective transmitter area.

Optical radiation propagates in vacuum at maximum speed c (300,000 km/s, natural constant) and in optically transparent media at lower speed v. The refractive index n = c/v is an optical material constant. Reflection is the reflection of light at mirror surfaces (angle of incidence equals angle of reflection). Refraction is the change in the direction of propagation at media interfaces with different refractive index (e.g. prism or lens surfaces). At interfaces air/glass, the light beam is directed towards the perpendicular, and at the glass/air interface it is directed from the perpendicular. In the latter case, depending on the refractive index combination, *total reflection* occurs, i.e. the light can no longer exit the medium with the higher refractive index and is forwarded, see Fig. 3.31. Total reflection is the basis of optical fibers (fiber core with high refractive index, fiber cladding with low refractive index), which allows light to be focused on "curved ways" can be guided. Applications: fiber optic sensors, endoscopes, pixelated image patterns.

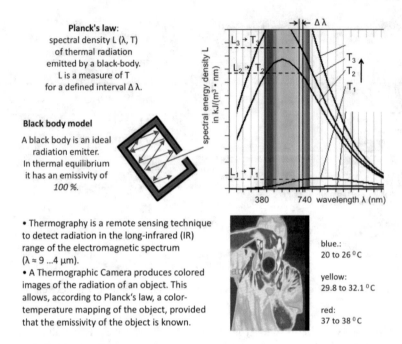

Planck's law:
spectral density L (λ, T)
of thermal radiation
emitted by a black-body.
L is a measure of T
for a defined interval Δ λ.

Black body model

A black body is an ideal
radiation emitter.
In thermal equilibrium
it has an emissivity of
100 %.

• Thermography is a remote sensing technique
to detect radiation in the long-infrared (IR)
range of the electromagnetic spectrum
(λ ≈ 9 …4 μm).
• A Thermographic Camera produces colored
images of the radiation of an object. This
allows, according to Planck's law, a color-
temperature mapping of the object, provided
that the emissivity of the object is known.

blue.:
20 to 26 °C

yellow:
29.8 to 32.1 °C

red:
37 to 38 °C

Fig. 3.30 Planck's law and its application in thermography

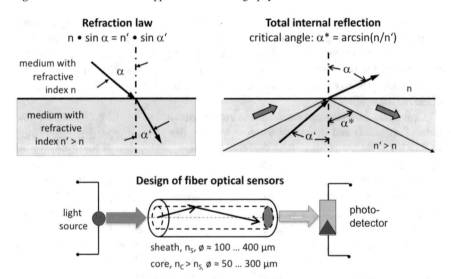

Refraction law
$$n \cdot \sin \alpha = n' \cdot \sin \alpha'$$

Total internal reflection
critical angle: $\alpha^* = \arcsin(n/n')$

medium with
refractive
index n

medium with
refractive
index n' > n

n

n' > n

Design of fiber optical sensors

light
source

photo-
detector

sheath, n_S, ⌀ ≈ 100 … 400 μm
core, $n_C > n_S$, ⌀ ≈ 50 … 300 μm

Fig. 3.31 Illustration of total reflection and the principle of fiber optics

3.6.5 Electromagnetism

Scientific studies of magnetite, already known in antiquity, and the "contact electricity" discovered by Thales on amber (Greek electron) led to the physical fields of magnetism and electricity, summarized under the term electromagnetism.

Magnetism

Magnetism was historically described as the force of certain ores of iron and also observed by the alignment of compass needles in the north–south direction of the earth. Magnetism is based onto different physical effects:

- Magnetic moments are properties of elementary particles of matter:
  - Diamagnetism is due to the orbital angular momentum of electrons in atoms, it does not appear on the outside.
  - Paramagnetism is based on the presence of permanent magnetic dipoles, due to incompletely occupied electron shells.
  - Ferromagnetism is the property of certain materials (e.g. iron, nickel, cobalt) to orient their elementary magnets in submicroscopic domains (Weiss domains) parallel to each other. The domains themselves cause a static magnetic field, or they are attracted to the magnetic pole of an external magnetic field. Applications are permanent magnets or magnetic data storage devices.

- Magnetic fields are created by every movement of electric charges. Stationary magnetic fields (magnetostatics) are generated by direct current (DC) and run in closed tracks. Oscillating charges (alternating current AC) cause electromagnetic fields, their periodic changes are electromagnetic waves.
- Induction is the creation of an electric field by changing a magnetic field, symbolically: magnetic field change ↔induction ↔electric field.

Electricity

Electricity is the name given to all phenomena that can be described by resting (electrostatics) and moving (electrodynamics) electrical charges. The theoretical basis for the characterization of the electrical properties of materials is the *electron band model,* illustrated in a simplified manner in Fig. 3.32.

According to Bohr's simplified historic atom model, electrons of isolated atoms (for example in a gas) can be considered to orbit at various distances about their nuclei, illustrated schematically by different energy levels, Fig. 3.32 (left). These distinct energy levels, which are characteristic for isolated atoms, widen into energy bands when atoms approach each other and form a solid. Quantum mechanics postulates that electrons can only "reside" within these bands, but not in the areas outside of them. The highest level of electron filling within a band is called the Fermi energy E_F. As shown in Fig. 3.32 (right), the electron energy-band representation classifies the electrical properties of materials as follows:

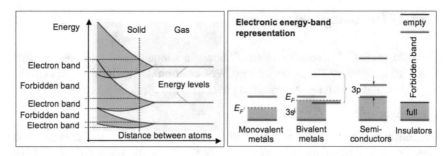

Fig. 3.32 Schematic representation of electron energy levels (left) and the electronic energy-band representation of materials (right)

- Monovalent metals (such as copper, silver, and gold) have partially filled bands. Their electron population density near the Fermi energy is high which results in a large conductivity.
- Bivalent metals are distinguished by overlapping upper bands and by a small electron concentration near the bottom of the valence band. As a consequence, the electron population near the Fermi energy is small which leads to a comparatively low conductivity. For alloys, the residual resistivity increases with increasing amount of solute content.
- Semiconductors have completely filled electron bands, but at elevated temperatures, the thermal energy causes some electrons to be excited from the valence band into the conduction band. They provide there some conduction. The electron "holes", which have been left behind in the valence band, cause a "hole current" which is directed in the opposite direction compared to the electron current. The number of electrons in the conduction band can be considerably increased (and thus the conductivity "tailored") by adding, for example to silicon small amounts of Group-V elements called donor atoms.
- Insulators have completely filled and empty electron bands which results in a virtually zero population density. Thus, the conductivity in insulators is virtually zero.

Electricity is characterized by the following basic operations and basic rules:

- In matter, there are uniformly distributed carriers of negative and positive electrical charges (electrons in the atomic shell, protons in the atomic nucleus). Charges of the same name exert a repulsive force on each other, and opposite polled charges exert a attracting force (Coulomb's law).
- In the vicinity of an electric charge there is an electric field with field lines, whose field strength E is defined by the local force F on a test charge Q: $F = Q \cdot E$. Locations of equal field strength lie on equipotential lines, the potential difference is the electric voltage U, measured in Volt V. The electric voltage is linked to the electric charge Q by the electric capacity C: $Q = C \cdot U$.

- Electric current is generated by the movement of electrical charge carriers (electrons, ions) in gases, liquids (electrolytes) and solids (metals, semiconductors). The amount of charge flowing per unit time is the electric current I, measured in amperes A.
- The electrical resistance R, measured in ohms, is a measure of the electrical voltage U required to cause an electric current I to flow. Ohm's law applies to electrical conductors: $U = R \cdot I$.
- The electrical energy is the product of charge and voltage: $E = Q \cdot U$ and the electrical power is the product of voltage and current: $P = U \cdot I$.
- In a stationary field of electric field strength E and magnetic flux density B, the Lorentz force $F = Q(E + v \times B)$ acts on an electric charge Q moving at the velocity v. The vectors v, B, F form a rectangular cartesian coordinate system. The Lorentz force can be used to convert electrical energy into mechanical energy (electric motor principle) and vice versa (generator principle).

The complete electromagnetism is described in closed form by the Maxwell equations. They can be formulated mathematically as differential equations or in integral form, their physical content is.

1. Electric charges are the sources of the electric field The electric flow through the closed surface of a volume is equal to the electric charge in its interior.
2. Every electric current is surrounded by closed magnetic field lines.
3. The field lines of magnetic flux density are always closed in themselves, i.e. there are no isolated magnetic poles.
4. Every time-varying magnetic field is surrounded by closed electric field lines (law of induction).

The solutions of Maxwell's equations describe electromagnetic waves whose existence was experimentally confirmed by Heinrich Hertz (1887). The totality of electromagnetic waves of different energies is called electromagnetic spectrum. It combines physical phenomena of electromagnetism and optics. Table 3.2 gives an overview with wavelength and frequency ranges as well as with technical applications.

Table. 3.2 Overview of the electromagnetic spectrum and application examples

Type of radiation	Wave length	Frequency	Application examples
Alternating current		50 Hz	Electricity supply
Ultrashort waves		30 … 300 MHz	Radio, television
Microwaves		3… 30 GHz	RADAR, satellite communication
Infrared radiation	1 … 780 nm		Heat radiation
Visible light	780… 380 nm		Illumination, lighting
Ultraviolet radiation	380 … 10 nm		Photochemistry
X-ray radiation	10… 10^{-3} nm	10^{16}… 10^{20} Hz	Diagnostics, materials testing
Gamma radiation	<1 nm		Radiotherapy

3.6.6 Quantum Theory

Quantum theory is based on the experimentally proven dualism of waves and parti-cles and contains Planck's quantum of action as fundamental natural constant. Three-dimensional space as the location of physical processes—relativistic connected with time—and the objects in space (particles, fields) are assumed to be empirically known. Quantum theory allows only probability predictions for measurements, the physical event is not fully causally determined. The measurement of the location of a quantum object is inevitably associated with a perturbation of its impulse and vice versa. The location x and the momentum p of a quantum object are always undeter-mined at least by amounts Δx and Δp and the product $\Delta x \cdot \Delta p$ cannot be smaller than Planck's quantum of action h. This is the **Heisenberg uncertainty relation:** $\Delta x \cdot \Delta p > h$. For the borderline case $h \to 0$, the corresponding classical theories follow from quantum theory, e.g. from quantum mechanics follows classical Newto-nian mechanics. Conversely, the classical theories of mechanics, electrodynamics and gravitation are transformed into the associated quantum theories with the help of quantization rules.

Quantum mechanics has its starting point in the analysis of the atomic structure of matter (see Sect. 3.1). It has been shown that an explanation of the atom as a "continuous electron space within the atomic shell" is impossible according to classical physics. Since, according to classical electrodynamics, the electrons must constantly emit light waves when moving on curved paths, they would eventually fall into the atomic nucleus. Bohr solved the problem with the "Bohr's atomic model". He applied Planck's "quantum hypothesis" to the atom and postulated that the possible "Electron orbits" are not a continuum, but a "discrete multiplicity", whereby there is a "deepest orbit", the "ground state" of the electron, from which the electron no longer radiates. Spectroscopic investigations (Franck-Hertz experiment) of the light emitted by excited atoms showed that in atoms, the energy absorption and release of energy is indeed only in the form of discrete energy packets, thus experimentally confirming the quantum theory. In the final version of quantum mechanics, however, electrons are not described by singular "orbits", but by the Schrödinger wave equation ψ. The wave function $\psi(x)$ is a complex function of the location x, whose absolute square $|\psi(x)|^2$ indicates the probability of finding the electron at location x.

Quantum electrodynamics (QED) is the relativistic quantum field theory of electr-dynamics. In essence, it describes how light and matter interact, and is the first theory where full agreement between quntum mechanics and special relativity is achieved. QED mathematically describes all phenomena involving electrically charged parti-cles interacting by means of exchange of photons and represents the quantum counter-part of classical electromagnetism giving a complete account of matter and light inter-action. Quantum electrodynamics was developed in essential features by Feynman which was awarded the Nobel Prize in 1965. He characterizes QED in his Lectures on Physics as follows:

> Quantum electrodynamics deals with the interaction between electrons and protons and is an electromagnetic theory which is correct in the sense of quantum mechanics. It is therefore

the fundamental theory of the interaction between light and matter or between electric field and charges. The theory contains rules for all physical phenomena and the known electrical, mechanical and chemical laws, with the exception of gravitation and nuclear processes.

Quantum gravity has not yet been experimentally confirmed. The concept represents an attempt to unite general relativity and quantum mechanics, see Sect. 3.2.

3.7 The World View of Physics

Physical observation, justification and experiment gave rise to today's knowledge of the world of physics. The physicist and philosopher Carl-Friedrich von Weizsäcker says: "Physics is the central discipline of natural science. Today we know no limits to its scope."

The *Standard Model of Cosmology* assumes that the universe arose about 14 billion years ago from a kind of "Big bbng" out of a space–time singularity. The physics of the Big Bang and the elementary forces are described in Sect. 3.3. Since then, the universe has expanded and cooled down, forming the structures we know today, from atoms to galaxies. If we start from the central concept of matter, the physical world is divided into the following **dimensional ranges of physics**:

- The nano world (dimension range; nanometers and below) with elementary particles, described by the Standard Model of Particle Physics,
- The micro world (dimensional range: nanometers to micrometers) with atoms (nucleons + electrons), described by nuclear physics and atomic physics,
- The macro world (dimension range: micrometers to meters and above) with gases, liquids and solids, described by the physics of matter,
- The universe (dimension range: light years) with celestial bodies and galaxies, explored by astrophysics. e.g. with the Hubble Space Telescope.

The experimentally observed physical phenomena in the different dimensions can be assigned to the following **basic areas of physics**:

- The first area is Mechanics, founded by Newton. It is suitable for describing all macroscopic mechanical processes, including the movement of liquids and elastic oscillations of bodies. Mechanics includes acoustics, aerodynamics and hydrodynamics. It also includes the astrophysics of the motion of celestial bodies.
- The second area is Heat, that developed from the insight that heat is not a substance but the energy of the movement of smallest particles, and can be described as Thermodynamics with methods of statistical mechanics. The concept of entropy is closely related to the concept of probability in the statistical interpretation of thermodynamics.
- The third area embraces Electricity, Magnetism, Optics and constitute the Electromagnetic Spectrum. Special Relativity and the theory of Matter Waves for elementary particles can also be included here. However, Schrödinger's wave

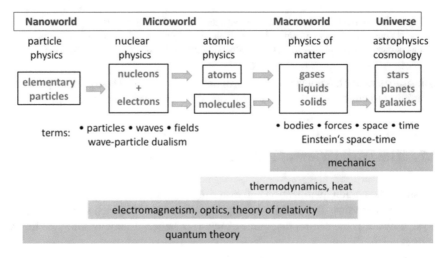

Fig. 3.33 The world view of physics

theory does not belong to this field; it is considered to be one of the foundations of quantum theory.

- The fourth area is Quantum Theory. Its central concept is the probability function, also known mathematically as the statistical matrix. This area includes Quantum Mechanics, Quantum Electrodynamics and the theory of Quantum Gravity, which, however, has not yet been experimentally confirmed.

The summary of the dimensional and conceptual areas can be described as the current world view of physics. It is shown with its facets in Fig. 3.33.

3.8 Antimatter: A Different World

Matter consists of atoms. They contain positive protons and neutral neutrons in the atomic nucleus and electrons with the negative elementary charge in the atomic shell. The existence of a "positron" as a positively charged "antiparticle" of the electron was theoretically postulated by Dirac in 1928 and experimentally proven by Anderson in 1932 in cosmic radiation. All properties of these two particles are the same, only the charges are reversed. The antiparticles of the other two stable constituents of matter, the antiproton and the antineutron, were artificially created in the high-energy accelerators in 1955 and 1956 respectively (Lawrence Berkeley National Laboratory). The further research work of high-energy physics and elementary particle physics showed that for every building block of our matter there exists a direct "antimatter partner". Antimatter is the collective term for antiparticles. When "normal" particles and the corresponding antiparticles meet, they can cancel each other out and their entire mass can be released as radiation energy.

The gate to another world

Under this headline, the Director General of the Research Center CERN, Rolf-Dieter Heuer, said to the Magazine SPIEGEL, July 9, 2014, the following:

Our so-called Standard Model of Cosmology describes only four to five percent of the universe. Almost a quarter accounts for "dark matter". We owe to this that the rotating galaxies are not traveling apart. This cannot be explained with the visible matter alone. The remaining almost three quarter fall upon what we call "dark energy". It causes that the universe expands even more rapidly. The "Higgs-field" which belongs to the "Higgs-particle" has a characteristic which fits to the dark energy: it acts equally in all directions. We found now the particle which awards all other particles their mass. Therewith, it is clarified that the Standard Model applies on the whole. Now, it is necessary to find the gap through which we cam advance to the remaining 95 % of the universe.

Fig. 3.34 A report on antimatter

The comparison of model calculations within the Standard Model of Cosmology (big bang theory) with astronomical measurement data suggests that the ratio matter and antimatter was almost 1 at the beginning. At exactly the same ratio matter and antimatter would have been completely converted into radiation during the cooling of the universe. A tiny imbalance—about 1 particle surplus to 1 billion particle/antiparticle pairs—caused a remnant of matter to remain, which is detectable in our universe today. This imbalance of matter and antimatter is one of the prerequisites for the stability of the universe and thus also for life on earth. The reason for this imbalance is one of the great mysteries of elementary particle physics and cosmology, see Fig. 3.34.

Chapter 4
Technology

Technology refers to the totality of man-made, benefit-oriented products and systems as well as the associated research, development, production and application. Technology in the 21st century is characterized by a combination of the "real physical world" of technical objects with the "virtual digital world" of information and communication. The chapter presents the essentials of techology and descibes the significance of technology for humanity.

Technology in the twenty-first century is marked by Friedrich Rapp in his book *Analytics for understanding the modern world* (Verlag Karl Alber 2012) as follows:

- *Modern technology has emerged from the combination of craftsmanship and skill and scientific method. Generally speaking, technology is about objects and processes of the physical world, which are created by social action based on the division of labor. Products can be manufactured and procedures applied that were previously completely unknown and there is hardly any area of life that is not shaped by modern technology.*

The development of technology is closely linked to the progress of physics. Figure 4.1 compares their basic areas in a simplified overview. Mechanics is the basis for all mechanical technologies from mechanical engineering to civil engineering and aeronautics. Energy technology is based on thermodynamics. Electromagnetism is the physical basis for electrical engineering and the use of "energy from the socket" for all areas of human life. Electronics enables today's fast information and communication technology and, with computer networks, the global Internet. Optics opens up the universe through telescopes and the microcosm through microscopes. Quantum physics is the physical basis for micro-technology and nano-technology.

Technology is based on physics, but the methodology of technology is fundamentally different from the methodology of physics.

- Physics explores nature with the methods of "reductionism and analysis", founded by Descartes and Newton: breaking down a problem into its simplest elements and their analysis for the recognition of elementary forces and natural constants.

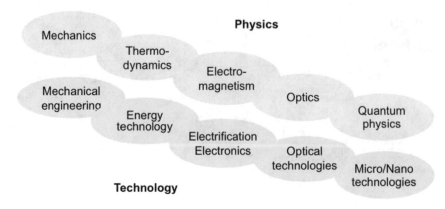

Fig. 4.1 Physics and technology in a comparative presentation of their fundamental areas

- In contrast, technology requires interdisciplinary engineering knowledge and a "holistic" methodology for the development, production and application of products and technical systems. Fundamental to technology are synthesis and systems thinking.

4.1 Dimensions of Technology

The dimensional ranges of today's technology cover more than twelve orders of magnitude. Figure 4.2 shows typical examples.

- Macrotechnology is the technology of machines, apparatus, equipment and technical installations: kilometer-long bridges and tunnels, pipelines, automobiles, ships and airplanes as well as all technical commodities.
- Microtechnology realizes today, with modules from micromechanics, microfluidics, micro optics, micro magnetics, microelectronics, the miniaturized functional components in high-tech devices. They range from computers, audio/video devices, smartphones and medical instruments up to the various household appliances.
- Nanotechnology has a dimensional range below micro technology (1 nm = 1/1000 μm). Nano-science was founded in 1960 by Richard P. Feynman (winner of the 1965 Nobel Prize in Physics). An examples of nanoscopic instrument technology is the atomic force microscope, the principle is depicted in Fig. 4.1. An atomic force microscopy (AFM) has a resolution more than 1000 times better than the optical diffraction limit. The information from an AFM is gathered by "feeling" or "touching" the surface with a piezoelectric actuator and displays the atomic surface topography.

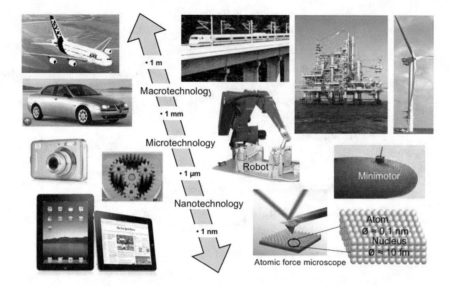

Fig. 4.2 Dimensions of today's technology, illustrated with typical examples

4.2 Fundamentals of Technology and Engineering

Engineering sciences are the sciences that deal with the research, development, design. Production, and application of technical products and technical systems. The classical engineering sciences are civil engineering, mechanical engineering, electrical engineering and production engineering. More recent courses of study are precision engineering, energy technology, chemical engineering, process engineering, environmental technology and technical computer science. In the twentieth century, new interdisciplinary courses of study were created in line with the growing needs of technology, business and society, e.g. mechatronics as a combination of *mechanics*—electronics—computer science or industrial engineering by combining economics (business administration, economics) and law with one or more engineering sciences to form a separate field of knowledge.

Based on the state of science and technology in the twenty-first century and the curricula of technical universities and colleges, the fundamentals of engineering sciences can be presented in a "knowledge circle" with four basic areas of disciplines, see Fig. 4.3:

1. Mathematical and scientific fundamentals

    Mathematics, Statistics, Physics, Chemistry.

2. Technological fundamentals

    Materials, Mechanics, Thermodynamics, Electronics, Metrology, Controls, Computer Engineering.

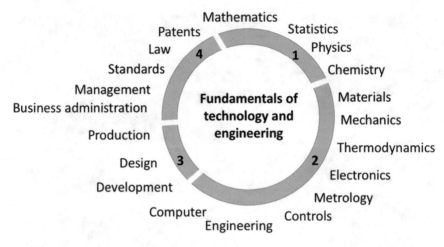

Fig. 4.3 The basic subjects of engineering sciences represented as "knowledge circle"

3. Fundamentals for products and services

   Development, design, production.

4. Economic and legal fundamentals

   Business administration, management, standardization, law, patents.

The individual disciplines required for interdisciplinary engineering sciences can be assembled from the knowledge circle in a modular fashion. For example, the collection of fundamentals for the curriculum of *Industrial Engineering* is illustrated in Fig. 4.4. The collection of fundamentals for the curriculum of *Mechatronics* is illustrated in Fig. 4.5.

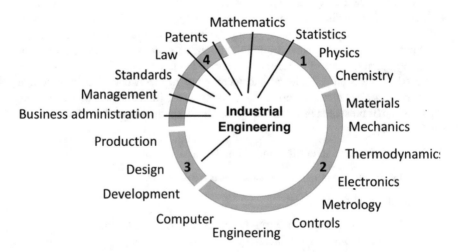

Fig. 4.4 Main disciplines for Industrial Engineering, selected from the circle of knowledge

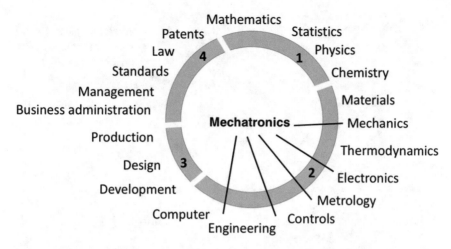

Fig. 4.5 Main disciplines for Mechatronics, selected from the knowledge circle

4.3 The Product Cycle and Base Technologies

The technology product cycle illustrates that products—accompanied by the necessary flow of energy, information, manpower and capital—move in "cycles" through the techno-economic system. From resources of matter and raw materials, the engineering materials—metals, polymers, ceramics, composites—are processed, and transformed by design and manufacture into products and systems. At the end of use, deposition or recycling of scrap and waste is necessary, Fig. 4.6.

The production cycle requires the succession of several technologies:

- Raw material technologies for the exploitation of natural resources.
- Material technologies for the production of materials and semi-finished products from the raw materials. The structural and functional materials, required for technical products, must correspond to the application profile and must accordingly be optimized.
- Design methods, i. e. the application of scientific, mathematical and art principles for efficient and economical structures, machines, processes, and systems.
- Production technologies, which concerns the manufacture of geometrically defined products with specified material properties. Means of production are plants, machines, devices, tools and other production equipment.
- Performance, maintenance and repair technologies to ensure the functionality and economic efficiency of the product.
- Finally, processing and recovery technologies to close the material cycle by recycling or, if this is not possible, by landfilling.

From an economic point of view, the production cycle can be regarded as an "economic value-added chain."

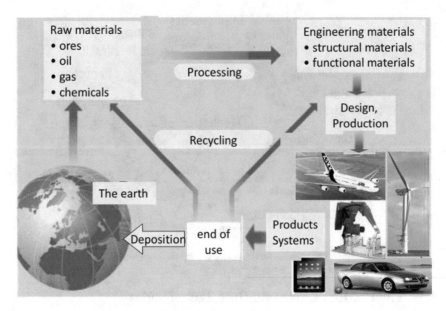

Fig. 4.6 Illustration of the technology product cycle

The illustration of the production cycle, Fig. 4.6, reminds us that the components of technical products and systems interact with their environment in their technical function. These interactions are generally described as two complementary processes:

- Immission, the effect of substances or radiation on materials and technical products, which can lead to corrosion, for example.
- Emission, the discharge of substances or radiation (also sound). An emission from a material is usually at the same time an immission into the environment.

To protect the environment—and thus people—there are legal regulations for emission and immission control with procedural regulations and limits for harmful substances and radiation. With regard to the protection of the environment, the main requirements are:

- Environmental compatibility, the property of not adversely affecting the environment in its technical function (and on the other hand not being adversely affected by the environment in question).
- Recyclability, the possibility of recovery and reprocessing after the intended use.
- Waste disposal, the possibility of disposing of material if recycling is not possible.

The technology product cycle, presented in Fig. 4.6, shows that the base technologies for all fabricated products are *Material, Energy and Information*.

4.3.1 Material

Materials can be natural (biological) in origin or synthetically processed and manufactured. According to their chemical nature, they are broadly grouped traditionally into inorganic and organic materials. Their physical structure can be crystalline or amorphous, as well as mixtures of both structures. Composites are combinations of materials assembled together to obtain properties superior to those of their single constituents. Composites are classified according to the nature of their matrix: metal, ceramic or polymer matrix composites, often designated as MMCs, CMCs and PMCs, respectively. Figure 4.7 illustrates, with characteristic examples, the spectrum of materials between the categories *natural*, *synthetic*, *inorganic*, and *organic*.

- Natural Materials: Natural materials used in engineering applications are classified into natural materials of mineral origin, e.g. marble, granite, sandstone, mica, sapphire, ruby, diamond, and those of organic origin, e.g. timber, India rubber, natural fibres, like cotton and wool. The properties of natural materials of mineral origin, as for example high hardness and good chemical durability, are determined by strong covalent and ionic bonds between their atomic or molecular constituents and stable crystal structures. Natural materials of organic origin often possess complex structures with direction-dependent properties. Advantageous application aspects of natural materials are recycling and sustainability.
- Ceramics (Inorganic non-metallic materials): Their atoms are held together by covalent and ionic bonding. As covalent and ionic bonding energies are much higher than metallic bonds, inorganic non-metallic materials, like ceramics have high hardness and high melting temperatures. These materials are basically brittle and not ductile: In contrast to the metallic bond model, a displacement of atomistic dimensions theoretically already breaks localised covalent bonds or transforms

Fig. 4.7 Classification of materials

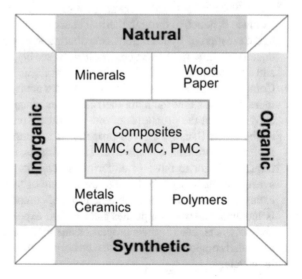

anion-cation attractions into anion-anion or cation-cation repulsions. Because of missing free valence electrons, inorganic non-metallic materials are poor conductors for electricity and heat, this qualifies them as good insulators in engineering applications.

- Polymers (Organic materials and blends): Organic materials whose technologically most important representatives are the polymers, consist of macromolecules containing carbon (C) covalently bonded with itself and with elements of low atom numbers (e.g. H, N, O, S). Intimate mechanical mixtures of several polymers are called blends. In thermoplastic materials, the molecular chains have long linear structures and are held together by (weak) intermolecular (van der Waals) bonds, leading to low melting temperatures. In thermosetting materials, the chains are connected in a network structure and do not melt. Amorphous polymer structures (e.g. polystyrene) are transparent, whereas the crystalline polymers are translucent or opaque. The low density of polymers gives them a good strength-to-weight ratio and makes them competitive with metals in structural engineering applications.

- Metals (Pure metals and alloys): In metals, the grains as the buildings blocks are held together by the *electron gas*. The free valence electrons of the electron gas account for the high electrical and thermal conductivity, and the optical gloss of metals. The metallic bonding—seen as interaction between the total of atomic nuclei and the electron gas—is not significantly influenced by a displacement of atoms. This is the reason for the good ductility and formability of metals. Metals and metallic alloys are the most important group of the so-called *structural materials* (see below) whose special features for engineering applications are their mechanical properties, e.g. strength and toughness.

- Semiconductors: Semiconductors have an intermediate position between metals and inorganic non-metallic materials. Their most important representatives are the elements silicon and germanium, possessing covalent bonding and diamond structure and the similarly structured III-V-compounds, like gallium arsenide (GaAs). Being electric non-conductors at absolute zero temperature, semiconductors can be made conductive through thermal energy input or atomic doping which leads to the creation of free electrons contributing to electrical conductivity. Semiconductors are important *functional materials* (see below) for electronic components and applications.

- Composites: Generally speaking, composites are hybrid creations made of two or more materials that maintain their identities when combined. The materials are chosen so that the properties of one constituent enhance the deficient properties of the other. Usually, a given property of a composite lies between the values for each constituent, but not always. Sometimes, the property of a composite is clearly superior to those of either of the constituents. The potential for such a synergy is one reason for the interest in composites for high-performance applications. However, because manufacturing of composites involves many steps and is labour intensive, composites may be too expensive to compete with metals and polymers, even if their properties are superior. In high-tech applications of advanced composites, it should also be borne in mind that they are usually difficult to recycle.

- Biomaterials: Biomaterials can be broadly defined as the class of materials suitable for biomedical applications. They may be synthetically derived from non-biological or even inorganic materials or they may originate in living tissues. The products that incorporate biomaterials are extremely varied and include artificial organs; bio-chemical sensors; disposable materials and commodities; drug-delivery systems; dental, plastic surgery, ear and ophthalmological devices; orthopaedic replacements; wound management aids; and packaging materials for biomedical and hygienic uses. For the application of biomaterials, the understanding of the interactions between synthetic substrates and biological tissues is of crucial importance to meet the needs of clinical requirements.

Scale of materials

The geometric length scale of materials has more than twelve orders of magnitude. The scale spectrum embraces nanoscale materials with quantum structures, microscale materials, and materials with macroscale architectures for macro engineering, assembled structures and engineered systems. Figure 4.8 illustrates the dimensional scale relevant for today's technology:

Processing of materials

For their applications, materials have to be *engineered* by processing and manufacture in order to fulfil their purpose as constituents of products, designed for the needs of economy and society. There are the following main technologies to transform matter into engineered materials, see Fig. 4.9:

- Net forming of suitable matter (liquids, molds), e.g. LASER printing.
- Machining of solids, i.e. shaping, cutting, drilling, etc.,
- Nano-technological assembly of atoms or molecules.

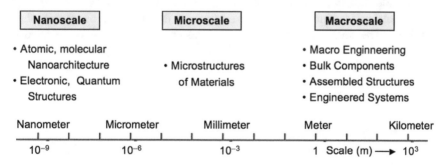

Fig. 4.8 Scale of material dimensions

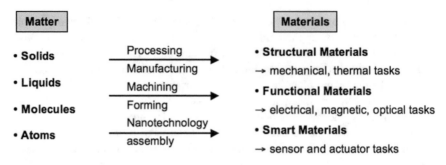

Fig. 4.9 Materials and their characteristics result from the processing of matter

Characteristics of materials

The consideration of the product cycle and the brief review of materials scales and processing methods show that the characterization of materials has to recognize many facets, see Fig. 4.10.

Materials result from the processing and synthesis of matter, based on chemistry, solid state and surface physics. Whenever a material is being created, developed, or produced, the properties or phenomena the material exhibits are of central concern. Experience shows that the *properties* and *performance* associated with a material are intimately related to its *composition* and *structure* at all scale levels, and influenced also by the engineering component design and production technologies. The final material—as constituent of an engineered component—must perform a given task under functional loads and environmental influences and must do so in an economical and societal acceptable manner.

Central to the characterization of materials is an approach connecting the "four elements" of materials science, namely: (1) synthesis and processing, (2) composition and microstructure, (3) properties, (4) performance. These fundamental features of materials, and the interrelationship among them, define the main materials characteristics. The whole of the four basic materials characteristics can be depicted as tetrahedron to indicate that they are inevitably connected, see Fig. 4.11.

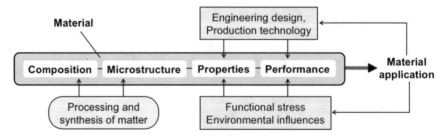

Fig. 4.10 The facets of materials

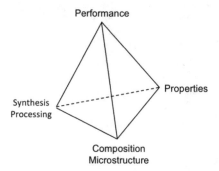

Composition and microstructure are *intrinsic (inherent)* materials characteristics. They result from the processing and synthesis of matter. The determination of these characteristics have to be backed up by suitable reference materials.

Properties and performance are *extrinsic (procedural) characteristics,* influenced by design and production. They describe the response of materials to functioanal stress and environmental influences on material's integrity.

Fig. 4.11   The basic four elements of materials characterization

The properties of materials, which are of fundamental importance for their engineering applications, can be categorised in three basic groups, broadly classified in the following categoriesategories

- **Structural materials**: engineered materials which have specific mechanical or thermal properties in responding to an external loading by a mechanical or thermal action.
- **Functional materials**: engineered materials which have specific electrical, magnetic or optical properties in responding to an external loading by an electromagnetic or an optical action.
- **Smart materials**: engineered materials with intrinsic or embedded *sensor* and *actuator functions,* which are able to accommodate materials in response to external loading, with the aim of optimizing material behavior according to given requirements for materials performance.

Numerical values for the various materials properties can vary over several orders of magnitude for the different material types. An overview on the broad numerical spectra of some mechanical, electrical and thermal properties of metals, inorganics and organics are shown in Fig. 4.12.

It must be emphasized that the numerical ranking of materials in Fig. 4.12 is based on "rough average values" only. Precise data of materials properties require the specification of various influencing factors described above and symbolically expressed as follows:

$$\text{Materials properties data} = f(\text{composition} - \text{microstructurescale; external loading; ...})$$

Application of materials for technological innovations and new industries

Engineering materials developed in the twentieth century are the basis for innovative industrial and economic sectors, here are a few examples:

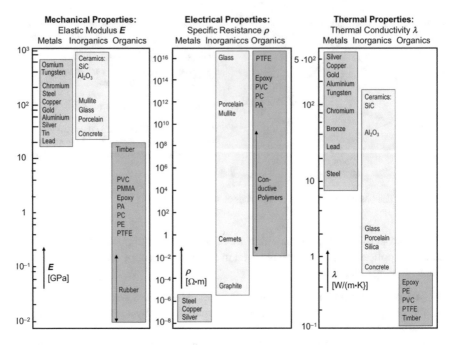

Fig. 4.12 Overview of mechanical, electrical, and thermal materials properties for the basic types of materials (metal, inorganic, or organic)

- Alumina alloys—developed in the **1920s** with 1/3 of the weight of Steel—enable lightweight construction, aircrafts, the aviation industry and global air traffic.
- Carbides have been increasing the productivity of industrial manufacturing technology since the **1930s** through a large increase in cutting speeds and tool life.
- Since the **1940s**, polymers have created important branches of industry in plastics technology; carbon fiber reinforced polymers are, for example, stronger than steel and only 1/5 of the weight.
- Since the **1950s**, super alloys have led to advanced developments in stationary and mobile turbine and jet engines with considerable increases of thermal efficiencies for power engineering and aircraft technology.
- Since the **1960s**, semiconductors have been based on new electronic, magnetic and photonic components the electronics, communication and information industry.
- Since the **1970s**, high-performance ceramics have been the basis for structural and functional Building blocks for further developments in many "high-tech areas".
- Biomaterials with improved biocompatibility were introduced in the **1980s** in the medical technology as implant material or organ replacement is becoming increasingly important.
- In the **1990s**, nanomaterials were interesting material developments with diverse technical application potentials in micro- and nanotechnology.

- Touch-screen materials with tactile-electronic signal conversion form the base for information and communication technologies in the digital world of the **twenty-first century**.

4.3.2 Energy

Energy is one of the fundamentals areas of physics and is divided into the following categories:

- Mechanical energy
  - Potential mechanical energy characterizes the working capacity of a physical system, which is determined by its position in a force field.
  - Kinetic mechanical energy is the work capacity that an object can generate due to of his movement. It corresponds to the work that must be done to move the object from rest to momentary motion. It depends on the mass and the velocity of the moving body.
- Electrical energy is defined as the working capacity that is generated by means of electricity, transmitted or stored in electric fields. Energy that is transferred between electrical energy and other forms of energy means electrical work.
- Thermal energy (heat energy) is involved in the disordered motion of atomic or molecular components of a substance. It is a state variable and is part of the inner energy. A radiation field has thermal energy when its energy is distributed disorderly among the various possible waveforms.
- Chemical energy is distributed in the form of a chemical compound, stored in an energy carrier and in chemical reactions which can be released.

All forms of energy are largely convertible into each other and therefore equivalent to each other. During energy conversions, the total energy, i.e. the sum of all energies of a closed system is constant (law of conservation of energy). **Energy is the working capacity of physical systems**. Thus, it is of outstanding technical and economic importance.

The **metrological unit of energy** (or work) is the joule (J). Because of the diversity of forms of energy and the different value ranges of released energy, different units have developed, which are equivalent to each other. Mechanical energy is often given in Newton meters (N m), thermal energy and chemical energy in general in the SI unit joule (J), electrical and magnetic energy in watt seconds (W s). The following equivalence applies: $1\,J = 1\,N\,m = 1\,W\,s$.

Energy technology is an interdisciplinary engineering science, with the technologies for the production, conversion, storage and use of energy. Primary energy production refers to the first stage where energy enters the supply chain before any further conversion or transformation process. Energy carriers are usually classified as

- Fossil, using coal, crude oil and natural gas,

- Nuclear, using uranium,
- Renewable, using hydro power, wind and solar energy, biomass, among others.

The energy content of energy sources is either heat that can be used for household and industrial purposes, or the energy content is made available by mechanical, thermal, or chemical transformation as electrical energy (electricity production). In a power station, mechanical energy is usually generated by means of generators into electrical energy, which is usually fed into the power grid. Mechanical (kinetic) energy for drive of generators comes from water or wind movements or uses—via steam turbines or gas turbines—thermal energy from burning coal, oil, natural gas, biomass, garbage or nuclear energy. Photovoltaic systems convert radiation energy directly into electricity. Important framework conditions for energy technology are

(a) The availability of energy sources,
(b) Environmental aspects (e.g. pollutant emissions),
(c) Energy conversion efficiency, the ratio of useful energy to primary energy, expressed as efficiency

$$\eta = E_{output}/E_{input} \text{ in percent.}$$

Examples of energy conversion efficiency:

- Heat production:

    - Gas heating 80 … 90%
    - Coal-fired furnace (industry) 80 … 90%
    - Solar collector up to 85
    - Electric range (household) 50 … 60%.

- Power generation:

    - Hydropower plant 80 … 90%
    - Gas/steam turbine power plant (natural gas) 50 … 60
    - Coal-fired power station 25 … 50%
    - Wind turbine up to 50
    - Nuclear power plant 33%
    - Solar cell 5 … 25%.

Energy is an indispensable resource for human existence. Just to stay alive, a human being needs an average daily food intake of about 3000 kilocalories (kcal), which represents a food-related energy equivalent of about 1200 kilowatt hours (kWh) per year. The total consumption requirement of people to energy (heating, electricity, fuels, etc.) has been estimated to be about 15 times higher. With a world population of more 7 billion people, the annual global energy demand sums up to about 140,000 billion kilowatt hours (500×10^{18} J) primary energy.

In Germany, the primary energy demand accounts for approximately 2.8% of the world's energy demand, an overview is given in Fig. 4.13. The basic feature of electricity energy is shown in Fig. 4.14.

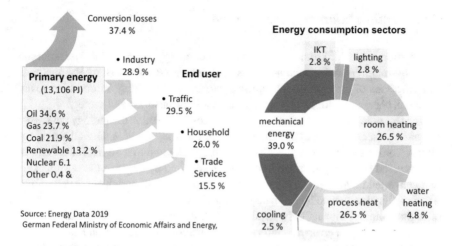

Fig. 4.13   Overview of energy technology: primary energy, end user and energy consumption sector

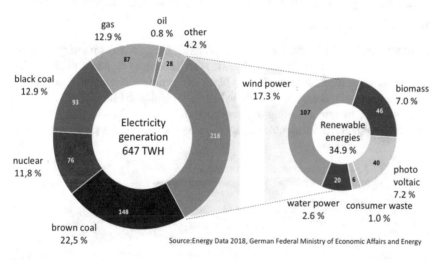

Fig. 4.14   The basic feature of electricity energy technology

   Electrical energy is the most versatile energy source, which can be converted into other forms of energy with very low losses. The availability of electric power is a basic prerequisite for any modern industry and cannot be replaced by other energy sources. Experience shows that a failure of the electricity supply brings any economy to a standstill and must therefore be limited as far as possible. A high level of security of supply of electricity is therefore an important condition for technology, economy and society.

4.3.3 Information and Digitalization

Information is the third basic technology besides energy and material. The umbrella term *information and communications technology (ICT)* refers to the methods and processes of generating, transmitting, storing and applying information (data, numbers, characters) for use by science, technology, business and society.

The technological bases of information and communication technology are *sensor technology* (see Sect. 3.5.2) and the *digitalization of information.* Digitalization (or digitization) is the conversion of text, pictures, or sound into a digital form that can be processed by a computer. An information, acquired by a sensor (sensor input) is first converted in an analog electrical signal (sensor output) and then converted into a binary signal sequence by an analog-to-digital AD converter. Since the processing of digital (binary) signals only requires a distinction between two signal states (0 or 1 or low or high), error-free information transfer to a computer and data storage is possible in principle. The information is usually transmitted in cryptographically secure information data blocks (block chains). Figure 4.15 provides an overview of the methodology of digitalization for information and communication technology.

Digitization of pictures

In order to digitize a picture, the picture is scanned, i.e. split into a matrix with rows and columns (the picture is "pixelated"). This can be done by scanners, photography, satellite-based or medical sensors. For a black and white raster graphic without grey tones, the image pixels are assigned the values 0 for black and 1 for white. The matrix is read out line by line, resulting in a sequence of the digits 0 and 1 that represents the picture. When digitizing color images, each color value of a pixel in the RGB

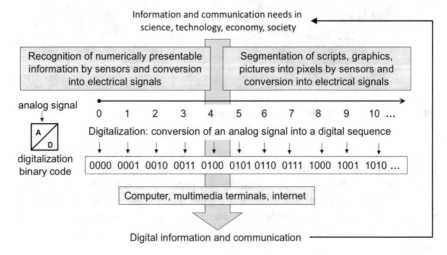

Fig. 4.15 The methodology of digitalization

(red-green-blue) color space is decomposed into the values red, green and blue, and these are stored individually with the associated quantization.

Text digitization

When digitizing text, the document is first digitized in the same way as an image, i.e. scanned. If the original appearance of the document is to be reproduced as accurately as possible, no further processing takes place and only the image of the text is stored. If the linguistic content of the documents is of interest, the digitized text image is translated back into a character set by a text recognition program, e.g. using the ASCII code (American Standard Code for Information Interchange) or, in the case of non-Latin characters, the Unicode (ISO 10646). Only the recognized text is then saved.

Digitization of audio/video signals

The digitization of audio data is called sampling. Sound waves are converted into analog electronic vibrations with a microphone as sensor and measured and stored as digital values. These values can also be "put together" to form an analog sound wave, which then can be made audible. Due to the large amounts of data that are generated, compression methods are used. This allow the realization of more space-saving data carriers, e.g. FLAC (Free Lossless Audio Codec) or MP3 (lossy compression of digitally stored audio data).

Optical data memories store data sequences in the form of pits and lands (plateau) in a compact disc (CD) for audio data or in a digital video disc (DVD) for video data. The data track information is read out in a CD/DVD player by mechanical contactless scanning with a LASER scanning unit when the CD/DVD is rotated. Pits and lands have a height difference leading to LASER-beam interference. The alternation of pits and lands is detected as a dark/light change (bit "1"). The laser beam reflection at individual pits or lands returns bits "0". This results in a serial 1-0 data stream (data output), which is fed to a photo detector, giving an output as an analogue audio or video signal sequence will.

4.4   Technical Systems

For the characterization of multicomponent technical products, instead of poorly distinguishable expressions such as "machine", "device", "apparatus", the generic term **technical system** has been fixed.

- *A technical system is characterized by its **function** to generate, convert, transport and/or store material, energy and/or information. It has a spatial **structure** and is composed of components, which emerge from design and production technology* (Brockhaus Encyclopedia, 2000).
- The conceptual base for technical systems is the **General Systems Theory**.

4.4.1 General Systems Theory

The classical method of scientifically characterizing objects of interest is "analytical reductionism". An entity, i.e. the object of an investigation, is virtually broken down into its individual parts so that each part could be analyzed separately, and the dissections could be added to describe the totality of the entity. This principle can be applied analytically in a variety of directions, e. g. resolution of causal relations into separate parts, searching for "atomic units" in science or for "material constants" in engineering. Application of the classical analytical procedure depends on the condition that interactions between the Parts are non-existent or, at least, weak enough to be neglected for certain research purposes.

The systems approach goes beyond the methodology of analytical reductionism. It considers how the parts interact with the other constituents of the system forming an entity of "organized complexity". In this approach, a system is a set of elements interconnected by structure and function. The behavior of a system is the manner in which the whole or parts of a system act and react to perform a function. In characterizing the behavior of systems, the terms structure and function also must not be isolated from each other, because the structure and the function of systems are interconnected. The basic features of systems can be summarized as follows:

- System definition: A system is a set of elements interconnected by structure and function

    (I)   Structure: $S = \{A, P, R\}$

    > A Elements(components)
    > $A = \{a_1, a_2, \ldots a_n\}n$: number of elements
    > P Properties of the elements
    > $P = \{P(a_i)\} \quad i = 1 \ldots n$
    > R Relations(interactions)between elements
    > $R = \{R(a_i \leftrightarrow a_j)\} \quad i, j = 1 \ldots n, j \neq i$

    (II) Inputs $\{X\}$, Outputs $\{Y\}$

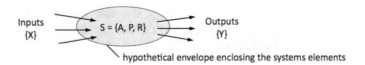

hypothetical envelope enclosing the systems elements

    (III) Function

    - Support of loads
    - Transfer or transformation of operating Inputs into functional outputs.

T Transfer function

- System behavior: the manner in which a system acts and reacts in performing a function.

4.4.2 Characteristics of Technical Systems

Emanating from the principles of General Systems Theory, the basic features of technical systems can be summarized as follows, Fig. 4.16

- A technical system is a set of elements (engineered components) interconnected by structure and function.
- The structure $S = \{A, P, R\}$ of technical systems consists of multiple components $\{A\}$, designed and manufactured with their properties $\{P\}$ and interactions $\{R\}$ for a given task.
- The function of technical systems is to generate, store, transmit or/and transform energy, materials and/or information. The function of a system is realized with operational inputs and functional outputs, supported by auxiliary inputs.

Technical systems are realized with **Systems engineering**: an approach to obtain a technical function with a structural ensemble of appropriate designed and interacting components.

4.4.3 Detrimental Influences on Technical Systems

In their applications, technical systems have to bear "functional loads" (stress) of mechanical, thermal and electromagnetic nature. They are also in contact with other solid bodies, aggressive gases, liquids or biological organisms because there is no usage of technical systems without interaction with the environment. All these actions on technical systems have to be recognized in order to avoid faults and failures, defined as follows:

- Fault (FR Panne, DE Fehlzustand): the condition of an item that occurs when one of its components or assemblies degrades or exhibits abnormal behavior.

Fig. 4.16 Characteristics of technical systems

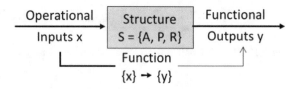

- Failure (FR Defailiance, DE Ausfall): the termination of the ability of an item to perform a required function. Failure is an event as distinguished from fault, which is a state.

The systematic examination of mechanisms which may cause faults and failures is called *root cause failure analysis*. *Reliability* is defined as the probability that a technical item will perform its required functions without failure for a specified time period (lifetime) when used under specified conditions.

An overview of the types of functional loads and environmental influences on technical systems and their detrimental influences is given in Fig. 4.17.

The detrimental influences on the structure and function of technical systems can be described briefly as follows:

- Fracture—the separation of a solid body into pieces under the action of mechanical stress—is the ultimate failure mode. It destroys the integrity and the structural functionality of materials and engineered components and must be avoided and prevented.
- Ageing results from all the irreversible physical and chemical processes that happen in a material during its service life. Thermodynamically, ageing is an inevitable process, however its rate ranges widely as a result of the different kinetics of the single reaction steps involved.
- Electromigration is the transport of material constituents caused by the gradual movement of ions in a conductor due to the momentum transfer between

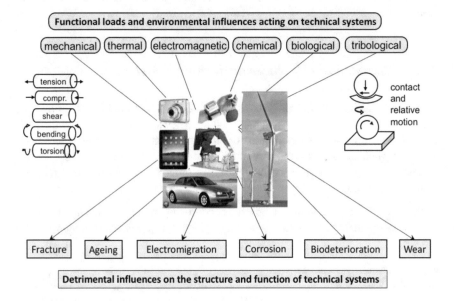

Fig. 4.17 Overview of functional loads and environmental influences on technical systems

conducting electrons and diffusing metal atoms. The effect is important in micro-electronics and related structures and can detrimentally influence the function of electronic materials.

- Corrosion is defined as an interaction between a metal and its environment that results in changes in the properties of the metal (ISO Standard 8044). In most cases the interaction between the metal and the environment is an electrochemical reaction where thermodynamic and kinetic considerations apply.
- Biodeterioration denotes detrimental changes of materials and their properties and may be caused by microorganisms (bacteria, algae, higher fungi, basidiomycetes) as well as by insects (termites, coleoptera, lepidoptera) and other animals.
- Wear is the process of deterioration of a solid surface, generally involving progressive loss of substance due to relative motion between contacting bodies. Wear is a complex process that depends on various system factors and interrelationships (see Sect. 4.5.2).

4.4.4 Structural Health Monitoring and Performance Control

Structural health monitoring involves the observation of a system over time using periodically sampled dynamic response measurements from an array of sensors, the extraction of damage-sensitive features from these measurements, and the statistical analysis of these features to determine the current state of system health. An example of the application of structural health monitoring is given in Fig. 4.18.

Performance control by condition monitoring is the control of a functional parameter of a technical system, such that a significant change is indicative of a developing fault. The application of conditional monitoring allows maintenance to be scheduled, or other actions to be taken to avoid the consequences of failure, before the failure occurs. Combining the methodologies of structural health monitoring and performance control, a general scheme to support the functionality of technical systems results, which is shown in Fig. 4.19.

4.4.5 Application of Technical Systems

The overviews of the dimensions of today's technologies (Fig. 4.2) and the product cycle (Fig. 4.6) illustrate the broad spectrum of technical systems. With regard to their functional areas and structural features, the following main categories of technical systems can be distinguished:

- Technical systems for the realization of motion and the transmission of mechanical energy and power, especially in mechanical engineering, production, and transport.

  → Tribological systems.

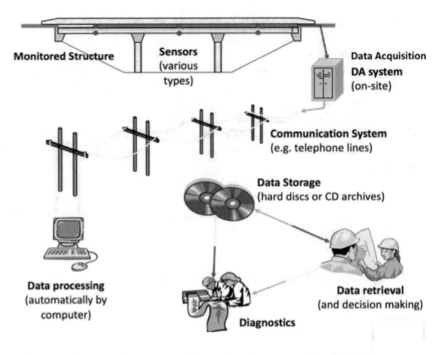

Fig. 4.18 Example of the monitoring and diagnostics of structures with embedded sensors

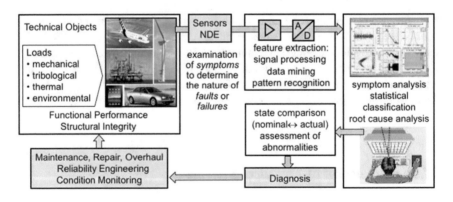

Fig. 4.19 A general scheme for the application of structural health monitoring and performance control to technical objects

- Technical systems that require for their tasks the combination of mechanics, electronics, controls, and computer engineering.

    → Mechatronic systems.

- Technical systems with a combination of mechatronics and internet communication

    → Cyber-physical systems.

- Technical systems for human health

    → Medical technology.

4.5 Tribological Systems

The term tribology was coined 1966 in the UK—after a comprehensive study on the enormous technological and economic importance of friction, lubrication and wear (Jost Report)—with the following definition: *Tribology is the science and technology of interacting surfaces in relative motion and of related subjects and practices.* The collection of controlling factors and relationships affecting friction and wear is referred to as a *tribological system*: a structure of interacting surfaces in relative motion to perform a specified technical function. Tribological systems have a "dual" character.

- On one hand, *"interacting surfaces in relative motion"* are necessary to realize technical functions in various technology branches like

    - *Mechanical engineering*: the transmission of motion and forces via interfacial traction
    - *Production*: the machining or forming of materials with workpiece-tool pairings
    - *Transport and traffic*: locomotion by wheel-rail or tire-road systems
    - Medical technologies: support of motion functions of the human body

- On the other hand, *interacting surfaces in relative motion* may lead to friction-induced energy dissipation and wear-induced materials deterioration. The impact of friction and wear on energy consumption, economic expenditure, and CO_2 emissions has been outlined as follows:

    - In total, about 23% of the world's total energy consumption originates from tribological contacts. Of that 20% is used to overcome friction and 3% is used to remanufacture worn parts and spare equipment due to wear and wear-related failures.
    - By taking advantage of the new surface, materials, and lubrication technologies for friction reduction and wear protection in vehicles, machinery and other equipment, energy losses due to friction and wear could considerably be reduced.
    - The largest energy savings through tribological expertise are envisioned in transportation (25%) and in the power generation (20%) while the potential savings in the manufacturing and residential sectors are estimated to be about 10%.

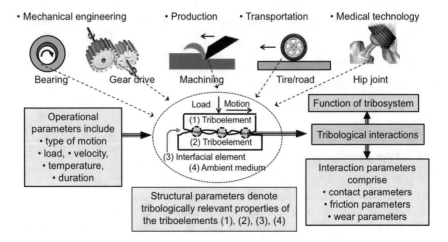

Fig. 4.20 Tribosystems in various technology areas and the basic groups of tribological parameters

4.5.1 Function of Tribological Systems

The science and technology of tribology is concerned with the optimization of technical systems from the study of friction and wear. This involves not only the properties of the materials in contact but also various system factors, such as the nature of the relative motion, the nature of the loading, the shape of the surface(s), the surface roughness, the ambient temperature, and the composition of the environment in which the wear occurs. As a consequence, some form of operational classification is needed to evaluate the material, mechanical, and environmental elements that can affect wear rates of a part. Typical tribological systems (in short tribosystems) are exemplified in Fig. 4.20. An overview of the broad range of the function of tribosystemes is compiled in Table 4.1.

4.5.2 Operational Parameters of Tribological Systems

The basic operational parameters of tribological systems include:

– Type of motion, classified in terms of sliding, rolling, spin, and impact and their possible superposition.
– The kinematics can be continuous, intermittent, reverse, or oscillating.
– Normal load (F_N), defined as the total force (including weight) that acts perpendicular to the contact area (A) between the contacting elements, where the contact pressure, p, is given by $p = F_N/A$.
– Velocity (v), specified as the vector components and the absolute values of the individual motions of the contacting elements.

Table 4.1 Function of tribosystems and typical examples

Function of tribosystems	Examples
Motion transmission and control	Bearings, Joints, Clutches, Cam and followers, Brakes
Forces and energy transmission	Gears, Rack-and-pinion, Screws, Drives, Actuators, Motors
Data and images storage and print	Computer hard disc drives, Printing units
Transportation	Drive and control technology,, Wheel/Rail, Tyre/Road
Fluid flow and control	Pipelines, Valves, Seals, Piston-cylinder assemblies
Mining	Dredging, Well drilling, Quarrying, Comminution
Forming	Casting, Drawing, Forging, Extrusion, Injection moulding
Machining	Cutting, Milling, Shaping, Boring, Grinding, Polishing
Medical technology	Medical Implants, Prosthetic devices, Dental technology

– Temperature (T) of the structural components at a stated location and time. In addition to the operating (steady-state) temperature, the friction-induced temperature rise (average temperature rise and flash temperatures), must be measured or estimated on the basis of friction heating calculations.
– Time dependence of the set of operational parameters (F_N, v, T), for example, load cycles and heating or cooling intervals.
– Duration (t) of the operation, performance or test.

4.5.3 Structure of Tribologicakl Systems

The structural parameters of a tribological system consist of the components in contact and relative motion with each other [triboelements (1) and (2)], the interfacial element (3) between the contacting parts, and the ambient medium (4). Examples of the four basic triboelemnts for the tribosystems shoen in Fig. 4.18 are given in Table 4.2.

The structure of tribosystems may be either "closed" or "open". Closed means that all components are involved in a continuous, repeated, or periodical interaction in the friction and wear process, for example, as in bearings or a gear drive. Open means that the element in the tribosystem is not continuously involved in the friction and wear process and that a materials flow in and out of the system occurs, for example, workpieces in machining.

Table 4.2 Examples of the structural elements of tribosystems

| Tribosystem | Triboelements | | Interfacial element (3) | Ambient medium (4) | System structure |
	(1)	(2)			
Bearing	Shaft	Bushing	Lubricant	Oil mist	Closed
Gear drive	Drive gear	Driven gear	Lubricant	Air	Closed
Machining	Tool	Workpiece	Cutting fluid	Air	Open
Tire/Road	Tire	Road	Moisture	Air	Open
Hip joint	Femur	Capsule	Synovial	Tissue	Closed

4.5.4 Interactions in Tribological Systems

Interaction parameters characterize the action of the operational parameters on the structural components of a tribological system. An overview is given in Fig. 4.21.

4.5.4.1   Contact Mechanics

The geometric configuration of contact between two parts is either conformal or counterformal (nonconforming) contact, Fig. 4.22. Depending on the configuration, the contact region may be a point, line, or area. The contact scales of tribological systems are illustrated in the lower part of Fig. 4.22.

- **Macroscale tribology** refers to tribosystems with dimensions from millimeters up to meters; sliding speeds are in the range of approximately 1 mm/s to more than 10 m/s. The function of macroscale engineering components—to transmit motion, forces, and mechanical energy—is governed by dynamic mass/spring/damping properties of the interacting elements.

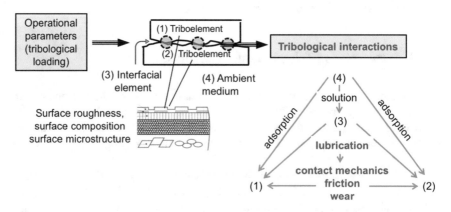

Fig. 4.21 Overview of interaction parameters of tribological systems

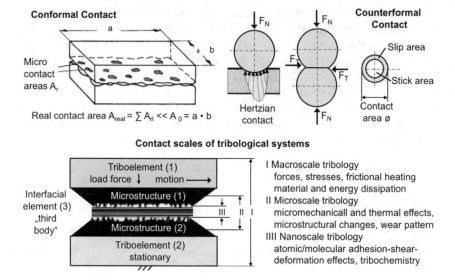

Fig. 4.22 Characteristics of tribological contacts

- **Microscale tribology** typically involves devices that are 100 to 1000 times smaller than their macroscale analogs. Thus, the volume of components on this scale is reduced by a factor of at least 10^6. As an example, the technology of microelectromechanical (MEM) systems is an interdisciplinary technology dealing with the design and manufacture of miniaturized machines, with the major dimensions at the scale of tens to hundreds of micrometers. Masses and inertias of MEM components rapidly become small as size decreases, whereas surface and tribological effects, which often depend on area, become increasingly important.
- **Nanoscale tribology** involves phenomena from the submicroscopic to the atomic scale. It is not possible to link the test results obtained at the nanoscale with friction and wear phenomena at the macroscopic scale, mainly because atomistic models are not directly scalable (Bullinger 2007). For the diagnosis of friction and wear test data it must be noted that the conditions in which nanotribology and macrotribology tests take place—using, for example atomic force microscopes (AFMs) for nanotribology and pin-on-disc testers for macrotribology—are very different. AFM tips induce stresses in the gigapascal range, whereas macroscale testers operate in the kilopascal range. Atomic force microscopes typically allow sliding amplitudes of only a few micrometers, whereas macrotesters sliding amplitudes range from hundreds of micrometers up to meters.

4.5.4.2 Friction

Solid friction is the resistance to motion when, under the action of an external force, one body moves or tends to move relative to another body. The quantity to be measured ("measurand" in the terminology of metrology) is the friction force, F_F, which is a vector quantity characterized by its direction and its quantity value. Depending on the kinematic conditions (i.e., sliding, rolling, spinning, or impact), the frictional force can be tangential that resists sliding or a frictional torque that resists rolling or spinning. The friction coefficient is defined as F_F/F_N. For sliding friction, the average frictional power P_F with a sliding velocity v is $P_F = F_F \cdot v$ (or $= f \cdot F_N \cdot v$). Frictional energy is $P_F \cdot$ duration.

The physics of solid friction has been explained by the Nobel laureate Richard P. Feynman in his famous Lectures on Physics as follows:

Physics of solid friction*

"Dry sliding friction occurs when one solid body slides on another. A force is needen to maintain motion. This force is called a frictional force and its origin is a very complicated matter. Both surfaces of contact are irregular, on the atomic level. There are many points of contact where the atoms seem to cling together, and then, as the sliding body is pulled along, the atoms snap apart and vibration ensues. As the slider snaps over the bumps, the bumps deform and then generate waves and atomic motions, and, after a while, friction-induced heat in the two bodies".

Sliding friction model on the atomic scale

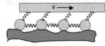

Tables with data of friction coefficients for materials, like "steel on steel" or "copper on copper", are all false. The friction is not due to "copper on copper" because the surfaces in contact are not pure copper but are mixtures of oxides and other impurities. It is impossible to get the right coefficient of friction for pure metals becaue if ultraclean pure metal surfaces are brought into contact the interfacial atomic forces became cohesive and the two pieces stick together. The friction coefficient which is ordinarily less than unity for reasonably hard surfaces becomes several times unity".

....

*   The Feynman Lectures on Physics, Addison.Wesley, 1963, Chapter 12-2 Friction.

The concise statement on the physics of solid friction was confirmed by measurements made by NASA for sliding friction of "copper on copper":

- $f = 0.08$, measured under boundary lubrication (mineral oil)
- $f = 1.0$, measured as solid friction in air
- $f > 100$ measured as solid friction in vacuum (10^{-10} Torr).

An overview of the mechanism of friction can be obtained from considering the energy balance (energy dissipation event) of solid friction The mechanical energy associated with solid friction between two solid triboelements (1), (2) involves the following, Fig. 4.23.

I.   Introduction of mechanical energy: formation of the real area of contact, junction growth at the onset of relative motion

II.   Transformation processes: (a) adhesion and shear, (b) plastic deformation, (c) abrasion, (d) hysteresis and damping

III. Dissipation processes: (i) thermal processes, (ii) absorption in (1) and (2), (iii) emission, e. g. heat, debris, noise.

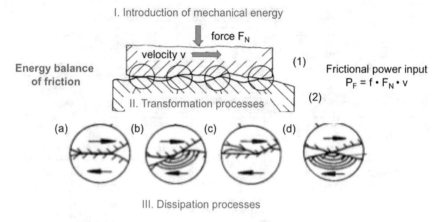

Fig. 4.23 Overview of the basic mechanisms of friction

4.5.4.3 Wear

Wear is the process of deterioration of a solid surface, generally involving progressive loss of substance due to relative motion between contacting bodies, that is, the interacting elements of a tribosystem. The chain of events leading to wear is illustrated in Fig. 4.24a. Tribological loading comprises the actions of contact mechanics, relative motion, and contacting matter, activated by frictional energy. The action of contact forces and stresses in combination with relative motion trigger the wear mechanisms of surface fatigue, abrasion, and the materials degradation processes listed in the left part of Fig. 4.24a. The contacting matter initiates—together with interfacial and ambient media—the wear mechanisms of tribochemical reactions and adhesion in combination with debris formation processes. All these processes and their interference lead to wear: surface damage and wear particles. The combined materials degradation effect of wear and corrosion is called tribocorrosion. The characteristic appearance of metallic surfaces worn by the basic wear mechanisms are shown in Fig. 4.24b.

4.5.4.4 Lubrication

Friction and wear can be significantly lower with fluid separation. The Stribeck curve characterizes the main friction and lubrication regimes as a function of the dimensionless film-thickness to roughness ratio λ as follows:

- Regime I, $\lambda < 1$: In this regime, friction is dominated by solid/solid interactions and shearing of interfacial boundary films.
- Regime II, $1 < \lambda < 3$: In this regime, friction involves a combination coexistence of solid-body friction (I) and fluid friction (III).

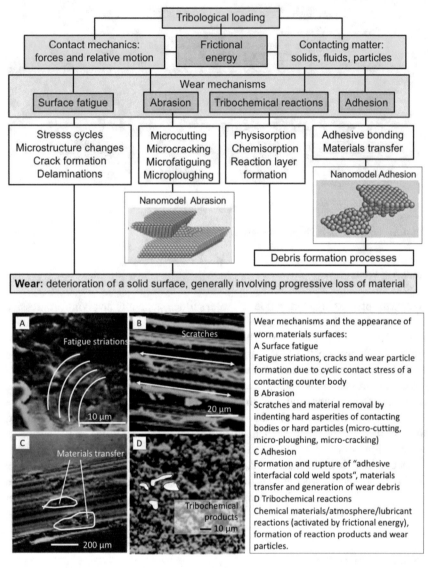

Figure 4.24 a The chain of events leading to wear **b** The appearance of worn surfaces (scanning electron microscopy, SEM)

- Regime III, $\lambda > 3$: In this regime, friction is dominated by lubricant rheology and hydrodynamics.

An overview of friction and lubrication regimes and the characteristic ranges of friction values is given on Fig. 4.25.

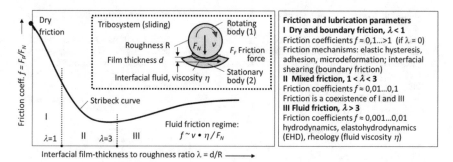

Fig. 4.25 The regimes of friction lubrication, characterized by the Stribeck curve

4.5.4.5 Tribotronics

The term tribotronics applies to the integration of electronic control in a tribological System. The electronic control network consists of a sensor, an actuator, and a control unit with real time software. The sensor detects a relevant tribomeasurand, for example friction force or interfacial displacement. The electrical sensor output is transferred via the controller unit to an actuator who act on the tribocontact in order to optimize the tribological system behavior. The concept of tribotronics is illustrated in Fig. 4.26. A comparison between a classical tribological system and a tribotronic system is made in Fig. 4.27.

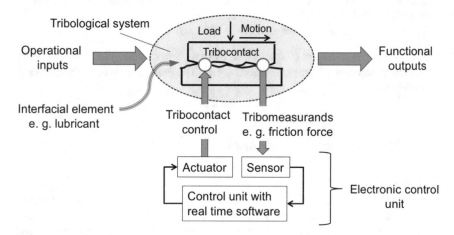

Fig. 4.26 Principle of tribotronic systems

Tribological System: Sliding bearing

Principle: Support of moving parts (load F_N, speed v) with friction. The friction force F_N is influenced by the interfacial element (lubricant) and characterized by the Stribeck curve.
I   Solid friction $f \approx 0{,}1...>1$,
II   Mixed friction $f \approx 0{,}01...0{,}1$
III Fluid friction $f \approx 0{,}001...0{,}01$

Tribotronic System: Magnetic bearing

Principle: Support of moving parts without physical contact by the electromagnetic levitation of actuators. The air gap s between the moving parts is kept constant at all loads and speeds with a controller-actuator unit. Friction is very low and wear is eliminated. Friction coefficient $f < 0{,}0001$

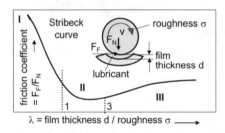

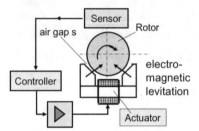

Fig. 4.27 Examples of tribological and tribotronic systems

4.6 Mechatronic Systems

The term *Mechatronics* was coined in the 1960s in Japan and is now globally acknowledged:

> Mechatronics is an interdisciplinary field of engineering, including mechanics, electronics, controls, and computer engineering.
>
> (Mechatronics, International Edition, McGraw-Hill 2003)

Mechatronic systems are characterized by their *structure* and *function*, according to the General Systems Theory (see Sect. 4.4.1). The general scheme of mechatronic systems is shown in Fig. 4.28.

4.6.1 Principles of Mechatronics

Mechatronic systems have a basic mechanical structure with multidisciplinary mechatronic modules and transform via the systems' structure operational inputs (energy, materials, information) into outputs needed to perform a technical function. The performance of mechatronic systems is controlled by sensors, processors and actuators. The principle of **sensors** together with a sensor selection matrix for sensor applications is described in Sect. 3.5. An **actuator** (in short actor) is a component in a mechatronic system that is responsible for the functional output by which the system acts to fulfil its task. An actuator requires a control signal and an electrical, fluidic or thermal power supply.

- Actuators with electric power supply

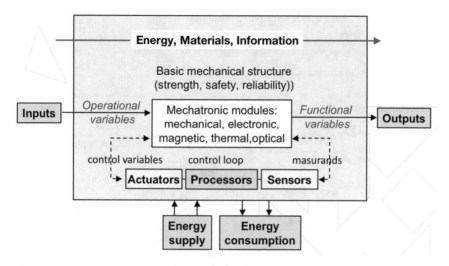

Fig. 4.28 General scheme of mechatronic systems

→ electric actuators, powered by a motor that converts electrical energy into mechanical.

→ galvanometer actuators produce a rotary deflection in an electro-magnetic field.

→ piezoelectric actuators, based on the inverse piezo-electrical effect.

- Actuators with fluidic power supply

    → hydraulic actuators, consisting of a fluid motor.

    → pneumatic actuators, converting energy formed by vacuum or compressed air.

- Actuators with thermal power supply

    → bimetallic strip that converts a temperature change into mechanical displacement.

    → shape-memory alloy, that switches between two crystal structures of different shape.

The principles of actuators and the different types are illustrated in Fig. 4.29.

The functional performance of mechatronic systems is controlled with a combination of actuators and sensors. The general block diagram for the control loop of mechatronic systems is shown in Fig. 4.30.

The major components of a mechatronic system control unit include a sensor, a controller and an actuator as final control element. The controller monitors the controlled process variable x, and compares it with the reference, the set point x_{SP}. The difference between actual and desired value of the process variable, called the controller error $e(t) = x_{SP} - x$, is applied as feedback to generate a control action to

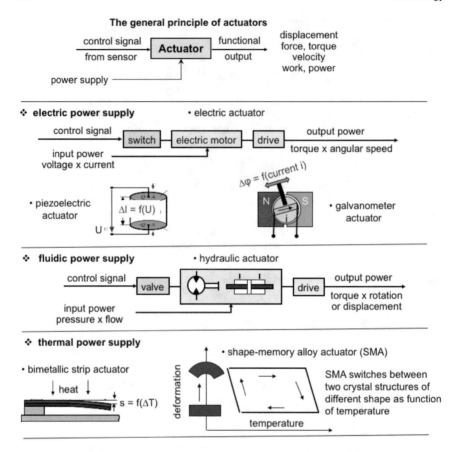

Fig. 4.29 The general principles and the different types of actuators

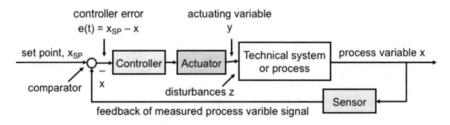

Figure 4.30 block diagram for the control of mechatronic systems

bring the controlled process variable to the same value as the set point. The diagram of Fig. 4.27 shows a closed loop system based on feedback. A closed loop controller has a feedback loop which ensures the controller exerts a control action to give a process output the same value as the "reference input" or "set point". To implement intermediate value control, the sensor has to measure the full range of the process variable.

4.6.2   Application of Mechatronics

Applied mechatronics is very broad. This has been expressed by the following statement

- *Virtually every newly designed engineering product is a mechatronic system.*

    Mechatronics, International Edition, McGraw-Hill 2003.

The application spectrum of mechatronic systems is illustrated in the following with examples from various technology areas:

- – industrial handling and production → robot
- – mobility → automobile mechatronics
- – information and communication → computer, smartphone
- – audio/video devices → CD player
- – photography → digital camera
- – typography → inkjet printer
- – Micro-mechatronics → MEMS, MOEMS
- – Nano-mechatronics → *Scanning tunneling microscope.*

4.6.2.1   Robot

Robots are a prominent category of mechatronic systems and indispensable today for industrial handling and production technologies. Industrial robots are mechatronic systems, instrumental for industrial handling and manufacturing. They are automated, programmable and capable of movement on two or more axes. Already in the year 2015, an estimated 1.63 million industrial robots were in operation worldwide according to the International Federation of Robotics (IFR).

An overview of the basic feature of an industrial robot is given in Fig. 4.31. On the left, the general description of the function and the structure is given, according to the general system theory (Sect. 4.4.1). A sketch of a robot as mechatronic system is shown on the right side of Fig. 4.31. The structure (S) of this mechatronic system is given by the robot components (A), their properties (P) and their motional inter-relation (R). The robot function is the transfer (T) of operational inputs (X) into functional outputs (Y). Sensors, processors and actuators form a control loop for the performance of the system.

The most commonly used robot configurations are Articulated Robots and SCARA robots, illustrated in Fig. 4.32. Typical applications of robots include pick and place actions from transport logistic to printed circuit step motions, welding, painting, assembly, product inspection, and testing, accomplished with high

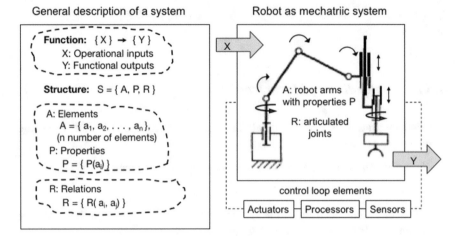

Fig. 4.31 The characterization of robots as mechatronic systems

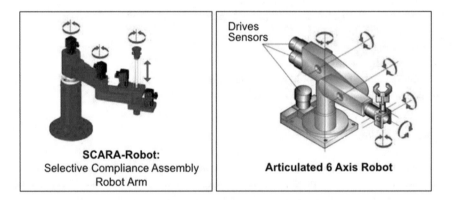

Fig. 4.32 Prominent types of industrial robots

endurance, speed, and precision. Instrumental for all these tasks is the mechatronic combination of mechanics, electronics, informatics and data mining with sensor technology.

4.6.2.2 Automobile Mechatronics

From a technological point of view, virtually all parameters relevant for road performance, safety and comfort of automobiles can be determined with appropriate sensors and controlled with mechatronic modules which include actuators and control algorithms. There is a broad variety of sensors which are suitable for application in an automobile. In Fig. 4.33, a selection of sensors for functional tasks in an automobile

Functional tasks and sensor examples

- steering wheel angle (AMR sensor)
- cam shaft (Hall sensor)
- crank shaft (Hall sensor)
- throttle (Hall sensor)
- anti-lock braking (inductive sensor)
- airbags (seismic sensor)
- tire pressure (micro mechanics sensor)
- temperature control (PTC sensor)

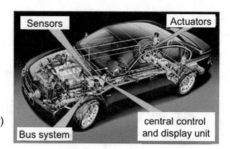

Fig. 4.33 Examples of mechatronic modules in an automobile

is shown. The sensors are part of a mechatronic control system consisting of sensors, actuators, a bus (bidirectional universal switch), and a central control and display unit.

Dynamic stability control (DSC), also referred to as electronic stability program (ESP), is a mechatronic technology that improves a car's driving stability by detecting and reducing loss of traction (skidding). DSC estimates with sensors the direction of the skid, and then applies via actuators the brakes to individual wheels asymmetrically in order to create torque about the vehicle's vertical axis, opposing the skid and bringing the vehicle back in line with the driver's commanded direction. During normal driving, the dynamic stability control works in the background and continuously monitors steering and vehicle direction. The principle of mechatronic dynamic stability control is explained in Fig. 4.34.

Driving stability is achieved with individual brake pulses to the outer front wheel to counter oversteer or to the inner rear wheel to counter understeer, see Fig. 4.35.

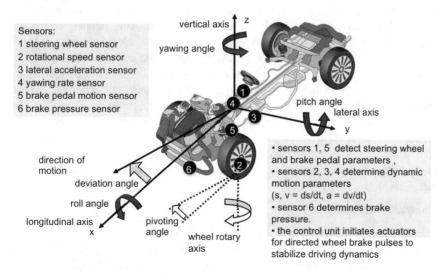

Fig. 4.34 Principle of dynamic stability control

Oversteer
yaw momentum clockwise

Understeer
yaw momentum anticlockwise

ESP-correction:
brake pulse to the outer front wheel,
correcting yaw momentum anticlockwise

ESP-correction:
brake pulse to the inner rear wheel,
correcting yaw momentum clockwise

Fig. 4.35 The control function of mechatronic dynamic stability control

4.6.2.3 Computer and Smartphone Mechatronics

Computers have a central role in information and communication technology. They are technical systems that can be instructed by an operator to carry out arithmetic or logical operations automatically via computer programming. Modern computers have the ability to follow generalized sets of operations, called programs. These programs enable computers to perform an extremely wide range of tasks.

A computer consists of a central processing unit (CPU), and a memory, typically a hard disc drive (HDD). The processing element carries out arithmetic and logical operations. Peripheral devices include input devices (keyboards, mice, etc.), output devices (monitor screens, printers, etc.), and input/output devices that perform both functions. Peripheral devices allow information to be retrieved from an external source and they enable the result of operations to be saved and retrieved. An overview of the computer's architecture together with a photo of a hard disc drive is given in Fig. 4.36.

Computer mechatronics

The design of a hard disc drive and its functional principle is depicted in Fig. 4.37 A typical hard disc drive has two actuators; a spindle motor that spins the disks and an actuator that positions the read/write head assembly across the spinning disks.

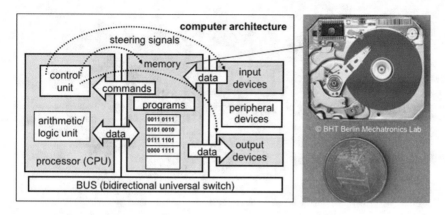

Fig. 4.36 The computer architecture illustrated by a block diagram and a photo of a hard disc drive

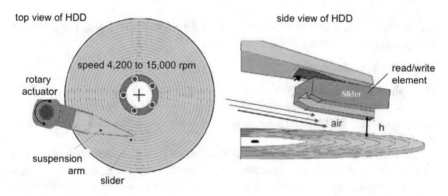

Fig. 4.37 The design of a hard disc drive and data of air gap dimensions and storage capacity

The actuator is composed of a permanent magnet and a moving coil often referred to as galvanometer actuator (see Fig. 4.29). In the start and stop position, the slider is parked outside the data track. In the operating state, the slider is separated from the rotating disc by an aerodynamically formed air gap that is responsible for storage density.

The physical principle of magnetic data storage in a HDD is illustrated in Fig. 4.38. A hard disc drive records data by magnetizing grains of typically 10 nm in a thin film of ferromagnetic material on a disk. The magnetic surface is divided into many small magnetic regions, referred to as magnetic domains. Changes in the direction of magnetization of the magnetic domains represent binary data bits. A change in the direction of their magnetization is associated with the logical level 1 while no change represents a logical 0. The data is read from the disk by detecting the transitions in magnetization. The head design techniques enabled a decrease of the air gap height

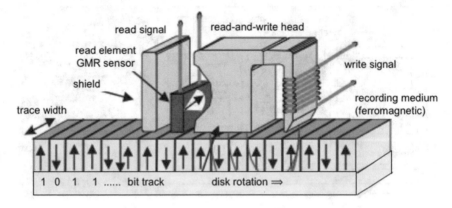

Fig. 4.38 Illustration of data storage and reading, and the principle of the GMR sensor in computer hard dis drives

down to 1... 2 nm, a data trace width of about 50 nm (500,000 traces per in.) and a storage capacity of about 1 Terabit/in.^{2}.

Smartphone mechatronics

Smartphones are multi-purpose information and communication devices. They are equipped with various mechatronic modules and sensors. An important sensor principle utilized for the tactile steering of smartphones is the capacitor principle, Fig. 4.39.

A capacitor consists of two conductors separated by a non-conductive region. A voltage V between the two conductors causes electric charges $\pm$ Q and an electric field. The capacitance C is defined by C = Q/V. The capacitance changes if the distance between the plates or the overlap of the plates changes.

- The capacity C of a distance sensor is inversely proportional to the input displacement d.
- The capacity of a displacement sensor is directly proportional to the overlap of the capacitor plates.

Capacitive **touchscreen sensors** consist of two transparent conductive traces of indium tin oxide (ITO), separated by a thin insulator, Fig. 4.40. The traces form a grid

Fig. 4.39 The principles of capacitor sensors

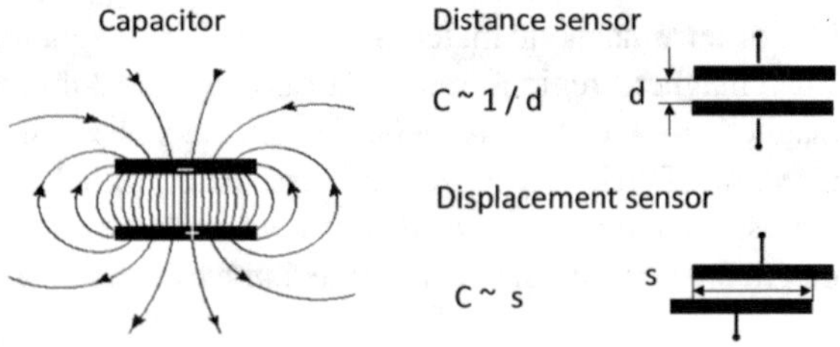

Capacitor

Distance sensor

$C \sim 1/d$ d

Displacement sensor

$C \sim s$ s

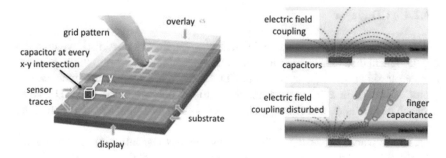

Fig. 4.40 The capacitor sensor in its application in a smartphone

pattern with x–y coordinates. At the x–y intersections local capacitors arise coupled by electric fields. A finger (or conductive stylus) near the surface of the touchscreen changes the local electric field and reduces the mutual capacitance. When a finger hits the screen, a tiny electrical charge is transferred to the finger creating a voltage drop at the contact point on the screen. The processing unit detects the x–y location of this voltage drop. This signal is used as steering input for the panel.

The application of a capacitor sensor for **image positioning** in smartphones is illustrated in Fig. 4.41.

Application example:
image positioning

(a) Smart phone vertical and sensor horizontal. No gravitational action on the moveable capacitor plate. i. e. $C_1 = C_2$, the display is vertical.
(b) Smart phone horizontal and sensor vertical.
The action of gravity on the moveable capacitor plate (mass m) leads to $C_1 \neq C_2$. The difference ΔC produces the steering signal to keep the display vertical.

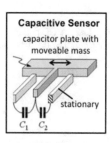

Capacitive Sensor

capacitor plate with moveable mass

stationary

C_1 C_2

Fig. 4.41 The capacitor sensor as applied for image positioning in a smartphone

4.6.2.4 CD Player

A **CD player** is a mechatronic device that plays audio compact discs which are a digital optical disc data storage format. A CD is made from 1.2 mm thick, polycarbonate plastic and weighs about 20 g. CD data is represented as tiny indentations ("pits") encoded in a spiral track molded into the polycarbonate layer. The areas between pits are known as "lands". The dimensions of a CD and a micrograph of the data track are shown in Fig. 4.42.

A CD is read by focusing a 780 nm wavelength (near infrared) semiconductor LASER housed within the CD player, through the bottom of the polycarbonate layer. The change in height between pits and lands results in a difference in the way the light is reflected. A change from pit to land or land to pit causes interference of a reflected LASER beam and indicates a bit "one", while no change indicates a series of "zero bits". Figure 4.43 shows the principle of data retrieval.

When a CD is loaded into the CD player, the data is read out according to the data retrieval principle explained in Fig. 4.43. An electric motor spins the disc with a variable rotary speed ω to obtain at all positions r a constant reading velocity v = $\omega \cdot$ r. The tracking control is done by analogue servo amplifiers and then the high frequency analogue signal read from the disc is digitized, processed and decoded into analogue audio signals. In addition, digital control data is used by the player to position the playback mechanism on the correct track, do the skip and seek functions and display track. An overview of the mechatronic modules needed for the functional performance of a CD player is given in Fig. 4.44.

Fig. 4.42 A CD and a micrograph of the data track

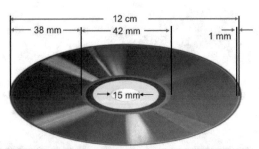

ca. 16,0000 windings, 6 km data length   CD thickness 1,2 mm

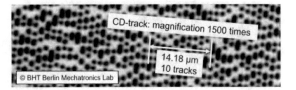

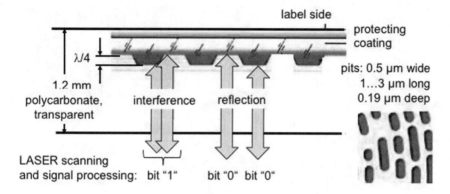

Fig. 4.43 The principle of the retrieval of data stored on a CD

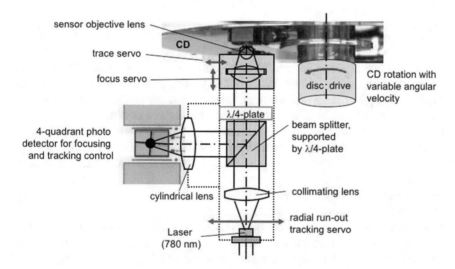

Fig. 4.44 The design of a CD player with the basic mechatronic modules

4.6.2.5 Digital Photography

Digital cameras are optical instruments used to record images. A camera captures light photons usually from the visual spectrum for human viewing, but in general could also be from other portions of the electromagnetic spectrum. All cameras use the same basic design: light enters an enclosed box through a lens system and an image is recorded on a light-sensitive medium (sensor). A shutter mechanism controls the length of time that light can enter the camera. A digital camera (or digicam) is a camera that encodes digital images and videos digitally and stores them for later reproduction. The technology for the digitization of images is explained in Sect. 4.3.

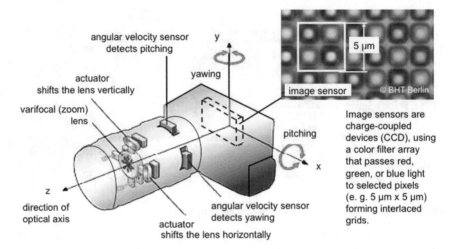

Fig. 4.45 Principle of optical image stabilization with MEMS and MOEMS in digital cameras

Key mechatronic modules are "Micro electro-mechanical systems" (MEMS) and "Micro opto-electro-mechanical systems" (MOEMS). They are responsible in digital cameras for optical image stabilization. Image stabilization (IS) is a technique that reduce blurring associated with the motion of a camera or other imaging device during exposure. Generally, it compensates for angular movement (yaw and pitch) of the imaging device. Mechatronic image stabilization can also compensate ±1° of rotation. The most common actuator is a galvanometer actuator (see Fig. 4.2). In combination with
strong permanent magnets, two coils are
used to drive a platform both vertically
and horizontally. As this system inherently creates a strong magnetic field, Hall positions sensors can be used to detect pitching and yawing. The actuators and position sensors can be tightly integrated in a small package, and work together in optical image stabilization, Fig. 4.45.

4.6.2.6 Inkjet Printing

Micro piezo actuators are the basic mechatronic modules in **inkjet printers**. They are a type of printing that recreates a digital image by propelling droplets of ink onto paper with the drop-on-demand technique, DOD. Piezo printers use an actuator nozzle array with a miniaturized ink-filled chamber behind each nozzle. The printing principle is explained in Fig. 4.46. In the suction phase (a), ink is taken into a chamber. In the rest phase (b), the chamber is filled with ink. When a voltage is applied, in the ejection phase (c), the piezoelectric actuator changes shape, generating a pressure pulse in the fluid, which forces a droplet of ink from the nozzle. A DOD process uses

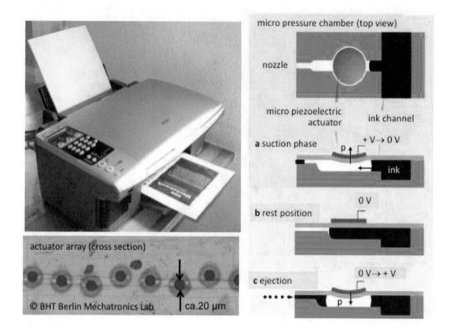

Fig. 4.46 The principle of inkjet printing with micro piezo actuators

software with an algorithm that directs the heads to apply between zero and eight droplets of ink per dot, only where needed.

4.6.3 Micro-mechatronics

Micro mechatronic systems are miniaturized mechatronic systems created by the methods and techniques of micro-technology. The terms "Micro electro-mechanical systems" (**MEMS**) and "Micro opto-electro-mechanical systems" (**MOEMS**) were coined for miniaturized items with moving parts. The general block diagram of micro mechatronic systems and the scaling requirements for sensors and actuators are shown in Fig. 4.47.

In microscale mechatronics, MEMS and MOEMS are miniaturized actuators. The design of different micro actuators for **MEMS** is illustrated in Fig. 4.48. Conventional electric motors operate through the interaction between an electric motor's magnetic field and winding currents to generate force or torque. They can be downsized to dimensions of about two millimeters, as shown on the left of Fig. 4.48. Actuators with dimensions in the micrometer range can be achieved with electrostatic motors, an example is shown in the middle of Fig. 4.48. On the right of Fig. 4.48, the characteristic data of piezo actuators are shown. They are based on the "inverse" piezoelectric effect.

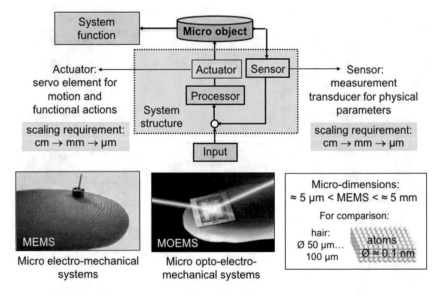

Fig. 4.47 General control-loop block diagram for micro mechatronic systems and the scaling requirements for sensors and actuators

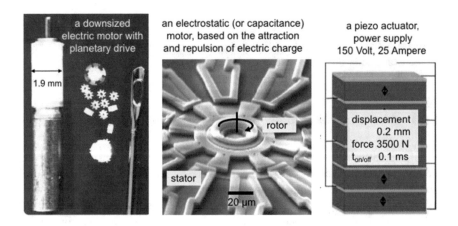

Fig. 4.48 Micro actuators for MEMS

Micro actuators for **MOEMS** are exemplified in Fig. 4.49 They utilize switchable electrodes and electrostatic motors for the optical transmission and display of signals and data.

Fig. 4.49 Micro actuators for MOEMs

4.6.4 Nano-mechatronics

Based on the development of miniaturized sensors and actuators, nanoscale mechatronic systems have been developed. Prominent examples of devices utilizing nanoscale mechatronics are the scanning tunneling microscope and computer hard disc drives.

The principle of the **Scanning Tunneling Microscopy** (STM) is explained in Fig. 4.50. The STM is based on the physical effect of quantum tunneling. When a

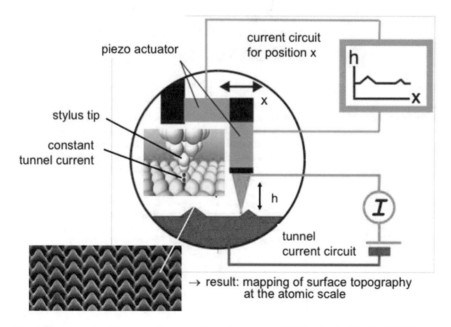

Fig. 4.50 Principle of the scanning tunneling microscope (Nobel prize in Physics 1986 for the inventors Binning and Rohrer)

conducting tip is brought very near to the surface to be examined, a voltage applied between tip and surface can allow electrons to tunnel through the vacuum between them. The resulting tunneling current is a function of tip position. A surface topography imaging on the atomic scale is acquired by monitoring the current as the tip's position scans with a piezo actuator across the surface, and is displayed in image form.

4.7   Cyber-Physical Systems

New technical systems were created at the beginning of the twenty-first century for which the term *Cyber-physical Systems (CPS)* was coined. The term is defined by the German Academy of the Technical Sciences ACATECH, (www.achatech.de) as follows:

- *Cyber-physical systems are characterized by a combination of real (physical) objects and processes with information-processing (virtual) objects and processes via open, partly global and at all times connectable information networks.*

An evolving cyber-physical systems landscape is anticipated for almost all areas of technology with the following examples:

Technology	Present	Future
Energy	Central generation, supervisory control and data acquisition systems for transmission and distribution	Systems for more efficient, safe, and secure generation, transmission, and distribution of electric power, integrated through smart grids. Smart ("net-zero energy") buildings for energy savings
Manufacturing	Computer-controlled machine tools and equipment. Robots performing repetitive tasks, fenced off from people	Smarter, more connected processes for agile and efficient production. Manufacturing robotics that work safely with people in shared spaces. Computer-guided printing or casting of composites
Materials	Predominantly conventional passive materials and structures	Emerging materials such as carbon fiber and polymers offer the potential to combine capability for electrical and/or optical functionality with important physical properties (strength, durability, disposability)

(continued)

(continued)

Technology	Present	Future
Transportation and mobility	Predominantly conventional passive materials and structures	Vehicle-to-vehicle communication for enhanced safety and convenience, drive-by-wire autonomous vehicles. Next generation air transportation systems
Medical care and health	Pacemakers, infusion pumps, medical delivery devices connected to the patient for life-critical functions	Life-supporting microdevices embedded in the human body. Wireless connectivity enabling body area sensor nets. Wearable sensors and benignly implantable devices. Configurable personalized medical devices

US Office of Science and Technology Policy: White Paper of Science and Technology for the 21st Century

Cyber-physical systems merge mechatronics with cybernetics, design and process science and communication. The Internet of things (IoT) is the inter-networking of physical devices, embedded with electronics, sensors, actuators, software, and network connectivity which enable these objects to collect and exchange data. CPSs are of relevance in such diverse industries as aerospace, automotive, energy, health-care, manufacturing, infrastructure, consumer electronics, and communications.

CPS apply RFID (radio-frequency identification) techniques to identify objects of interest and use sensors that collect physical data and act—with digital networks and actuators—on production, logistics and engineering processes, whereby they utilize multimodal man–machine interfaces. The general scheme of cyber-physical systems consisting of the combination of mechatronics with an internet combination is shown in Fig. 4.51.

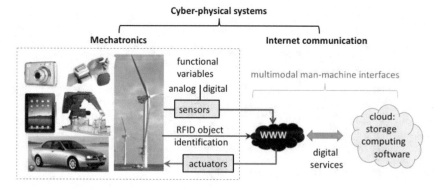

Fig. 4.51 The concept of cyber-physical systems

The most critical success factor for the realization and introduction of cyber-physical production systems is data, information and communication security. CPS networking creates security threats for industrial production with potential implications for functional safety aspects. Therefore, appropriate security architectures, protective measures and measures for the application of CPS and validation methods are required.

Industry 4.0

The term Industry 4.0 denotes a multidisciplinary technology, which evolved in the beginning of the twenty-first century, to apply the concept of cyber-physical systems (CPS) to industry. Central aspect of industry 4.0 is the ability of machines, devices, and people to connect and communicate with each other via the Internet of Things. Inter-connectivity allows operators to collect data and information from all points in the manufacturing process, thus aiding functionality. Industry 4.0 integrates processes in product development, manufacturing and service from suppliers to customers plus all key value chain partners.

With the help of cyber-physical systems that monitor physical processes, a virtual copy of the physical world can be designed. The technological conception of industry 4.0 and their essential parts are illustrated in Fig. 4.52

a. Characterization of the physical elements of industry 4.0: products, structures, systems, processes
b. Combination of physical elements to mechatronic systems with sensors, actuators, computers and control modules
c. Extension of mechatronic systems to cyber-physical systems (CPS) and internet communication
d. Cloud computing with available to many users over the Internet
e. Digital services with webservers or mobile applications.

Typical application areas of the concept of cyber-physical systems in industry 4.0 are

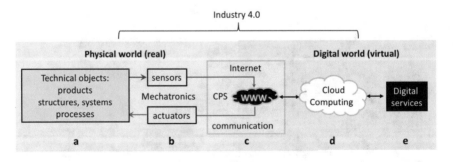

Fig. 4.52 The technological conception of Industry 4.0

- Power engineering with CPS technologies for linking decentralized generation and distribution of electrical energy, in order to achieve an optimal, need-based and stable to ensure the functioning of energy networks (smart grids).
- Transport technology with networking of vehicles ("Car-to-X") with each other or with the transport infrastructure via mobile radio. Cyber-physical systems will play a key role in future mobility, as they provide the basis for energy, battery and charging management.
- Product and production systems, which are controlled by CPS via component, plant, factory and company borders are networked with each other. This enables rapid production according to individual customer requirements. Also, the production process within companies can be made more adaptive, evolutionary and cooperative through a network of worldwide self-organizing production units of different operators.
- Medical technology systems, in which e.g. near-body sensors and medical information systems are connected via the Internet for remote monitoring. The acquisition of medical data by means of suitable sensor technology and their processing and evaluation in real time enables individual medical care.

4.8 Medical Technology

Medical technology is the application of the principles, methods and procedures of technology on humans and medicine. The basis for the applications of medical technology are body functions and bio signals of the body that can be recorded and therapeutically influenced,

A **biosignal** is any signal in living beings that can be continually measured and monitored, Fig. 4.53 gives an overview.

Biosignals characterize the medically relevant bodily functions of humans. They can be used with biosensor technology in systems diagnostics as functional variables (e.g. pressure, flow velocity, acoustic noise, temperature, electrical potentials) and as structural characteristics of humans (e.g. organ dimensions, volumes, elasticity, viscosity).

Biosignals are described—similar to signal functions in physics and technology— by signal shape, frequency, amplitude and the time of their occurrence. Their time response can be stationary, dynamic, periodic, discrete or stochastic. In medical technology applications, biosensors are combined with metrological components of signal processing and thus form a biological measurement chain, Fig. 4.54.

The technical systems of medical technology are medical devices: They are denoted in the EU *Directive on Medical Devices* (93/42/EEC) as follows: instruments, apparatus, substances or other objects which are used for the detection (diagnosis), prevention, surveillance (monitoring) and medical treatment (therapy) of human diseases or for the restoration of health (rehabilitation).

Medical devices are mechatronic systems. They consist of mechanics/electronics/IT components and work with sensor/processor/actuator

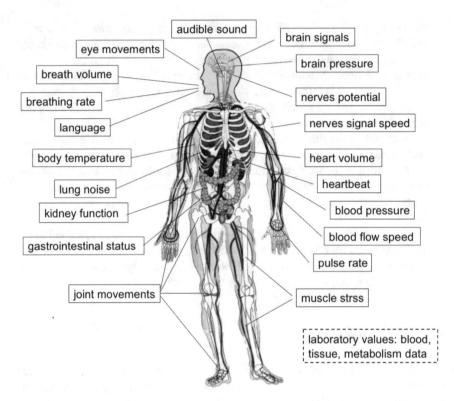

Fig. 4.53 Human body functions and the biosignals that characterize them

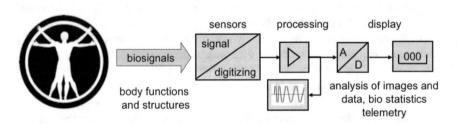

Fig. 4.54 Principle of a biosensor in a biological measurement chain

functional elements. In their medical technology applications by the physician they interact with the human being as a patient—especially in diagnostics, monitoring and rehabilitation. In technological terms, medical equipment technology is extremely diverse and is divided into two major areas:

- Body sensors for body functions, e.g. electro physical measurements of heart or brain functions, ECG, blood pressure sensors,

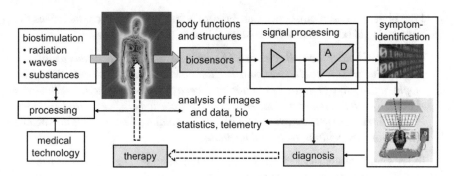

Fig. 4.55 Technology for humans: Biostimulation and biosensor technology for diagnosis and therapy

- Imaging procedures, e.g. X-ray diagnostics, sonography (ultrasound diagnostics), computed tomography, magnetic resonance imaging.

The application of medical technology by medical doctors is often associated with biotimulation, i.e. the application of stimuli, radiation, waves or medical substances to human test persons. This serves to improve the functionality of human sensory organs from the reaction to mechanical, acoustic, electrical, magnetic or visual stimuli. The determination of the resulting biosignals is carried out in a symptom analysis by biosensors using various methods:

- Sonography enables the analysis of ultrasound reflexes in the human body, the representation of movement sequences of heart valves and the measurement of blood flow velocity.
- X-ray absorption measurements allow the measurement of anatomical structures.
- By means of radioactively marked substances, metabolic phenomena and the transport speed, the accumulation site and the dynamics of precipitation processes can be determined.
- Ergonomic measurements can be used to determine the physical load capacity.

The principle of bioactuators and the interaction of biostimulation and biosensors is shown in a biological-mechatronic measuring chain in Fig. 4.55. The actual diagnostic findings are always provided by the responsible physician, supported by a variety of technological procedures and equipment.

4.9 Technology in the Twenty-First Century

"The Technological Age, which began at the end of the eighteenth century, is since the Axial Age (800-200 B.C.) the first mentally and materially completely new event", wrote 1949 the philosopher Karl Jaspers in his book *on the origin and aim of History*.

And the physicist and philosopher Carl Friedrich von Weizsäcker characterizes the technology in his book *Man in his History* (Hanser 1991) as follows:

- *Technology is older than all sciences. The early advanced civilizations begin with great achievements: Tools, road construction, town planning. The nutrition of the people in the cultural areas would not have been possible without the techniques of plough, wheel and cart. And how could people today (five billion in 1991, seven billion in 2018) be kept alive without technical agriculture, without means of transport and without technical medicine.*

Today's technology—based on findings in the natural sciences, new materials and engineering innovations—developed in four major phases from the middle of the eighteenth century onwards:

- **Period 1**: Industrial revolution through the development of mechanical technologies with the help from steam engines (James Watt 1769) and mechanical automation technology (e.g. loom 1785) → **Mechanization**
- **Period 2**: Electro-mechanics by combining mechanical and electrical technologies (electric motor, generator, Siemens 1866) → **Electrification**
- **Period 3**: Mechatronics through a systems engineering combination of electro-mechanics with electronics, computer engineering and information technology (1960s) → **Digitization**
- **Period 4**: Cyber-physical systems (CPS) in the combination of mechatronics and internet communication from 2000 on → **Internet of things**, → **Industry 4.0**.

The significance of technology for the world in the twenty-first century is described by Friedrich Rapp in his book *Analytics for understanding the modern world* (Verlag Karl Alber 2012):

- There is hardly any area of life that is not shaped by modern technology. However, our attitude towards technology is ambivalent. On the one hand, technology has become an integral part of our lives. Its radical rejection would endanger our physical existence and a substantial reduction of industrial technology would make at least the standard of living that we take for granted to a large extent impossible. On the other hand, however, the dangers of "over-mechanization" are also unmistakable. They range from resource consumption and the destruction of the natural environment to problems of data protection.
- The real problem is that everyone would like to enjoy the positive, welcome effects of technology, but according to St. Florian's principle—Saint Florian, spare my house, set fire to others!—no one is inclined to accept the inevitable negative side effects. Everyone wants the rubbish to be disposed of, and flight options to be available, but no one wants to live near a rubbish dump or an airfield.
- But modern technology, with its interconnected systems for energy transmission, transport and communication, is designed in such a way that the arrangements made are binding for all members of society. Since no one can boast of being in possession of absolute truth, the commandment of tolerance and pluralism is

essential for society—not only epistemological-philosophical but also practically-political.

A comprehensive analysis of the situation and significance of technology at end of the 20st century was conducted by the US National Academy of Engineering (www.nae.edu). The Greatest Engineering Achievements are shown in Fig. 4.56 in the order specified by the Academy.

The most important technology is ELECTRIFICATION. Its worldwide importance for all areas of human life as well as for all of technology, economy and society is obvious, and is clearly illustrated by the scenario of the catastrophic consequences of blackout: collapse of information and communication possibilities (telephone, TV, Internet) and failure of the entire electrified infrastructure, from rail transport to drinking water supply. Because of its central importance for technology, economy and society, ELECTRIFICATION is called the "Workhorse of the Modern World" by the Academy. The other major engineering achievements are disposed in four essential technology application groups in Fig. 4.57.

- The first group concerns the large and complex field of information and communication technologies, such as telephone, radio, television, electronics, computers and the Internet.

1. Electrification	11. Highways
2. Automobile	12. Spacecraft
3. Airplane	13. Internet
4. Water Supply and Distribution	14. Imaging
5. Electronics	15. Household Appliances
6. Radio and Television	16. Health Technologies
7. Agricultural Mechanization	17. Petrochemical Technologies
8. Computers	18. Lasers and Fiber Optics
9. Telephone	19. Nuclear Technologies
10. Air Conditioning and Refrigeration	20. High-performance Materials

Fig. 4.56 The Greatest Engineering Achievements

- Electrification

• Telephone	• Automobile	• Petrochemical Technologies	• Water Supply
• Radio and Television	• Airplane	• Nuclear Technologies	• Agriculture Mechanization
• Electronics	• Highways	• High-performance Materials	• Air Conditioning & Refrigeration
• Laser and Fiber Optics	• Spacecraft		• Household Appliances
• Computers			• Health Technologies
• Imaging	**Greatest Engineering Achievements**		
• Internet			

Fig, 4.57 The greatest engineering achievements, disposed in essential technology application areas

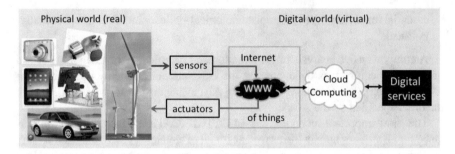

Fig. 4.58 Technology in the twenty-first century

- The second group names the most important technologies for human mobility, the automobile and the airplane.
- The third group concerns "enabling technologies", such as petrochemicals and high-performance materials.
- The fourth group includes technologies that are vital to the world's population of more than 7 billion people in the twenty-first century, from water supply technology and agricultural engineering to health technology.

New technical systems were created at the beginning of the twenty-first century for which the term *Cyber-physical Systems* was coined, see Sect. 4.7:

- *Cyber-physical systems are characterized by a combination of real (physical) objects and processes with information-processing (virtual) objects and processes via open, partly global and at all times connectable information networks.*

Cyber-physical Systems (CPS) are based upon the combination of technical objects with Internet communication (Internet of things). CPS utilize sensors for the detection of physical parameters of technical systems. The digital sensor outputs are used to steer and to control—via processors, actuators, suitable algorithms and cloud computing—production and engineering processes. CPS operate with multimodal interfaces and man-machine communication to provide universal "digital services" for the needs of industry and society. The duality of technology in the twenty-first century is illustrated in Fig. 4.58.

The significance of technology for humanity is described by the French philosopher Remi Brague in his book *The Wisdom of the World* (Beck 2006) as follows:

Technology is a kind of morality and maybe even the true morality. Today, technology is not only something that enables us to survive, it is more and more what enables us to live.

The Ancient Cosmos and the Triad Philosophy—Physics—Technology—Excerpt of the book

Excerpt of the book *The world is triangular*

Greek antiquity created with the concept of "Cosmos", the idea of a universal order of all things "in heaven and on earth". Philo-sophia was "love of knowledge", which also included nature (physis) and artistic and technical skills (techne). Later on, philosophy was understood purely in the humanities and physical and technical aspects were excluded from the scope of the term.

The model of the Cosmos is based on the "geocentric world view" with different "spheres":

- The *sublunar world* comprises the earth with perishable matter, plants, animals and humans, and extends with different zones (water, air, ether) up to the sphere of the moon.
- The *supralunar world* reaches to the border of the universe and knows no change, because it is divine in nature. The sun, the moon and all celestial bodies are a manifestation of the unseen divine.

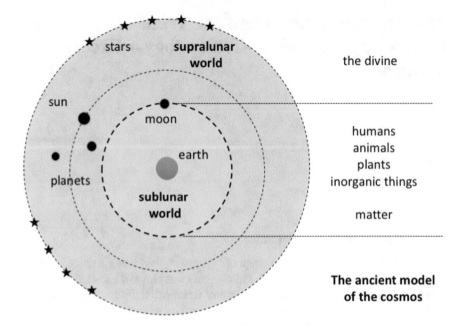

The ancient model of the cosmos

After the discovery of the real heliocentric world view by Copernicus, a "paradigm shift of world history" (Thomas Kuhn) took place: from the "closed cosmos model" to the "open world picture" of modern times with the triad *philosophy—physics—technology*.

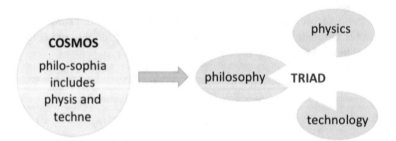

Philosophy

The world of philosophy is characterized by the relationship *man—nature—idea*. The basic directions of philosophy are symbolized with the Platonic Triangle.

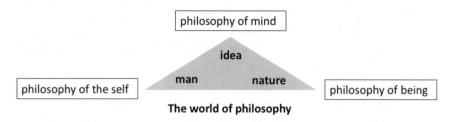

The world of philosophy

- *Philosophy of being*: The reflection on the world asks here for the "being" that underlies the observable phenomena. This is the approach of classical metaphysics, which today is called ontology (theory of being).
- *Philosophy of the self*: This direction of philosophical thinking starts with the "I"—also called "subject" in the language of philosophy. The main models are rationalism (Descartes, Leibniz, Spinoza) and empiricism (Locke, Hume, Berkeley). The combination of rationalism and empiricism was undertaken in the period of classical German idealism by Kant with his epistemology. A special variant of the philosophy of the self is the existential philosophy (Heidegger, Sartre).
- *Philosophy of mind*: Philosophizing here starts from the "idea" and develops philosophical models of the "absolute" in a synopsis of "being and I" (object and subject). These include the complex philosophical system of Hegel, materialism (Marx), the analytical philosophy (Russel, Wittgenstein) and the Three Worlds Theory (Popper).

The **Three-World-Theory** of Popper takes a mental division of the world into three interactive areas. Human psyche and consciousness acts as a mediator between physical reality and products of the human mind.

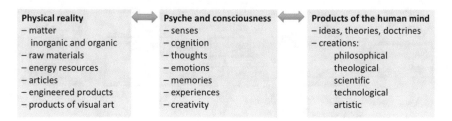

Physics

Physics explores and describes nature and the laws of nature. If one starts from the central concept of matter, the world of physics embraces four overlapping dimensional ranges, each with characteristic features of physical phenomena:

- Nanoworld, with elementary particles, described through particle physics,
- Microworld, described by nuclear physics and atomic physics,
- Macroworld, with gases, liquids and solids, described by the physics of matter,
- Universe with celestial bodies and galaxies, explored through astrophysics.

The Standard Model of Cosmology assumes that the universe was created about 13 billion years from a kind of "big bang". It has since then expanded, cooled and the structures we know today, from atoms to galaxies, were formed. The basic topical areas of physics are

- The first area is Mechanics, founded by Newton. It is suitable for describing all macroscopic mechanical processes, including the movement of liquids and elastic oscillations of bodies. Mechanics includes acoustics, aerodynamics and hydrodynamics. It also includes the astrophysics of the motion of celestial bodies.
- The second area is Heat, that developed from the insight that heat is not a substance but the energy of the movement of smallest particles, and can be described as Thermodynamics with methods of statistical mechanics. The concept of entropy is closely related to the concept of probability in the statistical interpretation of thermodynamics.
- The third area embraces Electricity, Magnetism, Optics and constitute the Electromagnetic Spectrum. Special Relativity and the theory of Matter Waves for elementary particles can also be included here. However, Schrödinger's wave theory does not belong to this field; it is considered to be one of the foundations of quantum theory.
- The fourth area is Quantum Theory. Its central concept is the probability function, also known mathematically as the statistical matrix. This area includes Quantum Mechanics, Quantum Electrodynamics and the theory of Quantum Gravity, which, however, has not yet been experimentally confirmed.

The summary of the dimensional and conceptual areas can be described as the current world view of physics.

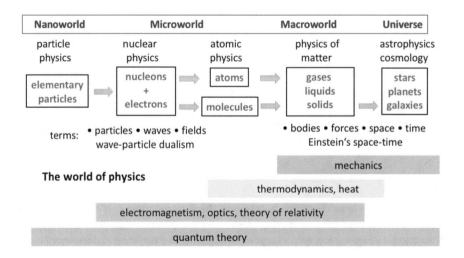

Technology

Technology denotes the totality of man-made useful products and systems from the macro-scale to the nano-scale.

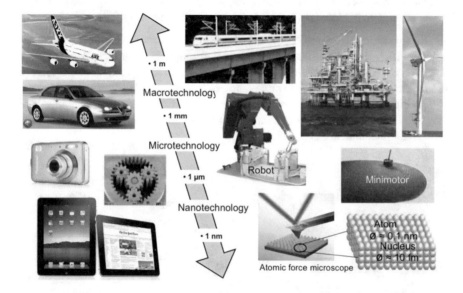

A comprehensive analysis of the situation and significance of technology at end of the 20st century was conducted by the US National Academy of Engineering (www.nae.edu) with the title "Greatest Engineering Achievements".

The most important technology worldwide is ELECTRIFICATION, denoted as "Workhorse of the Modern World". The other greatest engineering achievements of technology can be grouped in four essential technology areas.

- The first group covers the complex area of information and communication technologies.
- The second group comprises the technologies for human mobility.
- The third group calls "enabling technologies", such as petrochemical and high performance materials.
- The fourth group comprises technologies that are crucial to the world population, from water supply technology and agricultural technology to health technology.

	• Electrification		
• Telephone	• Automobile	• Petrochemical Technologies	• Water Supply
• Radio and Television	• Airplane	• Nuclear Technologies	• Agriculture Mechanization
• Electronics	• Highways	• High-performance Materials	• Air Conditioning & Refrigeration
• Laser and Fiber Optics	• Spacecraft		• Household Appliances
• Computers			• Health Technologies
• Imaging	**Greatest Engineering Achievements**		
• Internet			

New technical systems were created at the beginning of the 21st century for which the term *Cyber-physical Systems* was coined:

- *Cyber-physical systems are characterized by a combination of real (physical) objects and processes with information-processing (virtual) objects and processes via open, partly global and at all times connectable information networks.*

Cyber-physical Systems (CPS) are based upon the combination of technical objects with Internet communication (Internet of things). CPS utilize sensors for the detection of physical parameters of technical systems. The digital sensor outputs are used to steer and to control—via processors, actuators, suitable algorithms and cloud computing—production and engineering processes. CPS operate with multimodal interfaces and man-machine communication to provide universal "digital services" for the needs of industry and society.

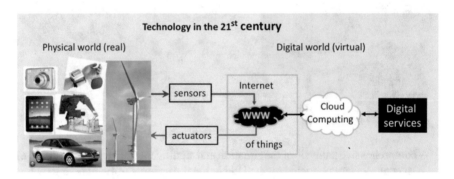

References

Anzenbacher, A.: Einführung in die Philosophie. Verlag Herder, Freiburg (2010)

Blackburn, S.: Denken–Die großen Fragen der Philosophie. Primus, Darmstadt (2001)

Brague, R.: Die Weisheit der Welt–Kosmos und Welterfahrung im westlichen Denken. Verlag C. H. Beck, München (2006)

Breitenstein, P.H., Rohbeck, J. (eds.): Philosophie–Geschichte, Disziplinen. Kompetenzen. Verlag J. B. Metzler, Stuttgart (2011)

Bullinger, H.-J. (ed.): Technologieführer–Grundlagen, Anwendungen, Trends. Springer Verlag, Berlin Heidelberg (2007)

Coogan, M.D. (ed.): Weltreligionen. Taschen, Köln (2006)

Czichos, H.: Measurement, Testing and Sensor Technology–Fundamentals and Application to Materials and Technical Systems. Springer, Berlin Heidelberg (2018)

Czichos, H.: Mechatronik Grundlagen und Anwendungen technischer Systeme, 4th edn. Springer Vieweg, Wiesbaden (2019)

Czichos, H. (ed.): Handbook of Technical Diagnostics. Springer, Berlin (2011)

Czichos, H., Habig, K.-H.: Tribologie-Handbuch, 4th edn. Springer Vieweg, Wiesbaden (2015)

Czichos, H., Hennecke, M. (eds.): HÜTTE–Das Ingenieurwissen. Springer Verlag, Berlin Heidelberg (2012)

Eliade, M., Couliano, I.P.: Das Handbuch der Weltreligionen. Patmos, Düsseldorf (2004)

Einstein, A.: Mein Weltbild. Ullstein Verlag, Berlin (1956)

Einstein, A., Infeld, L.: Die Evolution der Physik–Von Newton bis zur Quantentheorie. Rohwolt, Hamburg (1956)

Feynman, R.P., Leighton, R B., Sands, M.: The Feynman Lectures on Physics. Addison-Wesley Publishing Company, Reading, 1963

Feynman, R.P.: Was soll das alles? (Originaltitel: The meaning of it all). Piper Verlag, München (1999)

Feynman, R.P.: QUED Die seltsame Theorie des Lichtes und der Materie. Piper, München (2018)

Fischer, A.: Die sieben Weltreligionen. Sammüller Kreativ, Fränkisch-Crumbach (2006)

Gerthsen, Chr: Physik. Springer Verlag, Berlin Heidelberg (1960)

Gessmann, M.: Philosophisches Wörterbuch. Alfred Kröner Verlag, Stuttgart (2009)

Göbel, E.O., Siegner, U.: The New International System of Units (SI). Wiley-VCH, Weinheim (2019)

Grabner-Haider, A. (ed.): Philosophie der Weltkulturen. Marixverlag, Wiesbaden (2006)

Harari, Y.N.: Eine kurze Geschichte der Menschheit. DVA, München (2015)

Hawkings, S.: Kurze Antworten auf große Fragen. Klett-Cotta, Stuttgart (2018)

Heer, F.: Gottfried Wilhelm Leibniz. Fischer-Bücherei, Frankfurt am Main (1958)

Heer, F.: Die großen Dokumente der Weltgeschichte. Wolfgang Krüger Verlag, Frankfurt am Main (1978)

Heisenberg, W.: Das Naturbild der heutigen Physik. Rohwolt, Hamburg (1958)

Heisenberg, W.: Physik und Philosophie, 8th edn. Hirzel-Verlag, Stuttgart (2011)

Helferich, C.: Geschichte der Philosophie. dtv-Verlag, München (2009)

Hellmann, B.: Der kleine Taschenphilosoph. dtv-Verlag, München (2004)

Höffe, O.: Lesebuch zur Ethik. Verlag C. H. Beck, München (2012)

Honerkamp, J.: Wissenschaft und Weltbilder. Springer Verlag, Berlin Heidelberg (2015)

Jaspers, K.: Die maßgebenden Menschen–Sokrates, Buddha, Konfuzius, Jesus. Piper Verlag, München (2015)

Jaspers, K.: Vom Ursprung und Ziel der Geschichte. München & Zürich, 1949. Neuauflage Schwabe Verlag, Basel (2016)

Jordan, S., Mojsisch, B. (eds.): Philosophen Lexikon. Philip Reclam, Stuttgart (2009)

Jordan, S., und Nimtz, C. (Hrsg.): Lexikon Philosophie. Philip Reclam, Stuttgart (2009)

Jost Report: Tribology–A Report on the Present Position and Industry's Needs. Department of Education and Science, Her Majesty's Stationery Office, London, UK (1966)

Kaiser, K., König, W.: Geschichte des Ingenieurs. Carl Hanser Verlag, München (2006)

Kirchhoff, T., Karafyllis, N.C. (eds.): Naturphilosophie. Mohr Siebeck, Tübingen (2017)

Küng, H.: Der Anfang aller Dinge–Naturwissenschaft und Religion. Piper Verlag, München (2007)

Kunzmann, P., Burkhard, F.-P.: dtv-Atlas Philosophie. Deutscher Taschenbuch Verlag, München (2011)

Köhler, M.: Vom Urknall zum Cyberspace. Wiley-VCH Verlag, Weinheim (2009)

Lichtenegger, K.: Schlüsselkonzepte zur Physik. Springer Verlag, Berlin Heidelberg (2015)

Magee, B.: The Story of Philosophy. Penguin, London (2016)

Martienssen, W., Röß. D. (Hrsg.): Physik im 21. Jahrhundert. Springer Verlag, Berlin Heidelberg (2011)

Nicola, U.: Bildatlas Philosophie–Die abendländische Ideengeschichte in Bildern. Parthas Verlag, Berlin (2007)

Poller, H.: Die Philosophen und ihre Kerngedanken. Olzog Verlag, München (2007)

Popper, K.: Die offene Gesellschaft und ihre Feinde. Teil 1 und Teil 2. Francke Verlag, München (1980)

Popper, K., Eccles, J.C.: Das Ich und sein Gehirn. Piper, München (1989)

Rapp, F.: Analysen zum Verständnis der modernen Welt: Wissenschaft–Metaphysik–Technik. Verlag Karl Albe Freiburg (2012)

Resag, J.: Feynman und die Physik. Springer Verlag, Berlin Heidelberg (2018)

Reti, L. (ed.): The Unknown Leonardo. McGraw.-Hill, London (1974)

Röd, W.: Der Weg der Philosophie. Band I Altertum, Mittelalter, Renaissance. Band II 17. bis 20. Jahrhundert. Verlag C. H. Beck, München (1996)

Römpp, G.: Aristoteles. Böhlau Verlag, Köln (2009)

Römpp, G.: Platon. Böhlau Verlag, Köln (2008)

Röthlein, B.: Sinne, Gedanken, Gefühle–Unser Gehirn wird entschlüsselt. dtv-Verlag, München (2004)

Rovelli, C.: Sieben kurze Lektionen über Physik. Rowohlt Verlag, Hamburg (2016a)

Rovelli, C.: Die Wirklichkeit, die nicht so ist, wie sie aussieht. Rowohlt Verlag, Hamburg (2016b)

Scheibe, E.: Die Philosophie der Physiker. Verlag C. H. Beck, München (2006)

Schrödinger, E.: Die Natur und die Griechen–Kosmos und Physik. Rohwolt Verlag, Hamburg (1956)

Schulte, G.: Philosophie. Dumont, Köln (2006)

Schwanitz, D.: Bildung. Eichborn, Frankfurt am Main (2002)

Simonyi, K.: Kulturgeschichte der Physik. Urania Verlag, Leipzig (1990)

Spur, G.: Technologie und Management–Zum Selbstverständnis der Technikwissenschaft. Carl Hanser Verlag, München (1998)

von Weizsäcker, C.-F.: Der Mensch in seiner Geschichte. Carl Hanser Verlag, München (1991)

von Weizsäcker, C.-F.: Zeit und Wissen. Carl Hanser Verlag, München (1992)

Printed in the United States
by Baker & Taylor Publisher Services